M000316392

# SCIENCE GIANTS
## LIFE SCIENCE

**25 Activities Exploring the World's Greatest Scientific Discoveries**

ALAN TICOTSKY

Aligns to National Science Education Standards

A GOOD YEAR BOOK™

Good Year Books
Tucson, Arizona

# Dedication

*To Judy, Dave, Matthew, and Adam,*
*with much love and gratitude.*

---

*Science Giants: Life Science* contains lessons and activities that reinforce and develop skills defined in the National Science Education Standards developed by the National Research Council as appropriate for students in grades 5 to 8. These include diversity and adaptations of organisms, structure and function in living systems, reproduction and heredity, regulation and behavior, and populations and ecosystems. In addition, the following areas of the standards are central to the approach of *Science Giants:* unifying concepts and processes, science as inquiry, and the history and nature of science. See www.goodyearbooks.com for information on how specific lessons correlate to specific standards.

---

**Good Year Books**

Our titles are available for most basic curriculum subjects plus many enrichment areas. For information on other Good Year Books and to place orders, contact your local bookseller or educational dealer, or visit our website at www.goodyearbooks.com. For a complete catalog, please contact:

Good Year Books
PO Box 91858
Tucson, AZ 85752-1858
www.goodyearbooks.com

Cover Design: Sean O'Neill
Text Design: Dan Miedaner
Drawings: Sean O'Neill
Cover Images: Left to right, top: Photograph of Charles Darwin, courtesy of ©Bettman/Corbis; Photograph of Rachel Carson, courtesy of U.S. Fish and Wildlife Service; Portrait of Gregor J. Mendel, ©Bettman/Corbis; bottom: Photograph of George Washington Carver, courtesy of Library of Congress; Photograph of Louis Pasteur, courtesy of Library of Congress.

ISBN-10: 1-59647-106-9
ISBN-13: 978-1-59647-106-1

1 2 3 4 5 6 7 8 9 - ML - 13 12 11 10 09 08 07 06

Library of Congress Cataloging-in-Publication Data

Ticotsky, Alan.
  Science giants : life science : 25 activities exploring the world's greatest scientific discoveries / Alan Ticotsky.
    p. cm.
  Includes bibliographical references (p.  ).
  ISBN-13: 978-1-59647-106-1
  ISBN-10: 1-59647-106-9
    1. Life sciences--History--Study and teaching (Middle school)--Activity programs. 2. Discoveries in science--Study and teaching (Middle school)--Activities programs. I. Title.

QH305.T63 2006
570.71'2--dc22
2006043509

# Contents

# Introduction for Teachers

Some ideas people believed in the past appear foolish to us. Other ideas seem to be inevitable but erroneous conclusions reached using limited resources and information. One may assume with reasonable certainty that some of today's prevailing knowledge will be overturned by new discoveries in the future. Science can describe reality to the limit of our tools and our ability to conceptualize that which we cannot measure or observe. New ideas constantly challenge old assumptions. Revolutions occur when an idea is discarded in favor of a better one.

## The Structure of This Book

*Science Giants: Life Science* arranges important scientific discoveries in major disciplines into a historical context. Activities and simulations provide hands-on experiences for students using readily available classroom supplies. Activities are followed by student reading pages summarizing the history of scientific discovery and explaining the theories.

The book is divided into chapters based on major areas of science inquiry. Each chapter contains teacher instructions for active student investigations paired with student reading pages. You can use chapters individually, or you can follow the sequence of the book to provide an overview of the history of life science.

Activities are designed for teams of students and follow a simple format—a list of materials needed per team (mostly common, inexpensive items), followed by instructions and teacher background information. Teamwork among students provides valuable rewards in the classroom. Working in teams:

- encourages dialogue among students, creating better thinking and more discovery.
- improves communication skills.
- increases motivation.
- promotes peer teaching and learning.
- builds social competency.

After doing an activity, hand out the student reading pages to enhance students' knowledge of the history behind each discovery. Student reading pages include vocabulary words, which are shown in bold type and defined at the end of each reading, and offer suggestions for further study. Each reading is self-contained and has all the necessary information for students to understand the significance of the scientific achievements. Time lines at the beginning of each chapter provide reference points and springboards for studying biography, an important and interesting aspect of the history of science. There's a bibliography for teachers at the end of the book.

The book focuses on ideas rather than personalities. Some famous legends are covered because scientific and historical literacy would be incomplete without them. The circumstances of discovery often illustrate the truth of Louis Pasteur's observation, ". . . chance favors the prepared mind." Using *Science Giants* should help prepare the minds of students for future discoveries.

## Gender Equity

Why is there a predominance of men in the history of science? Margaret Cavendish was the first woman elected to the prestigious Royal Society in 1667; the next woman member was admitted in 1945. Examples of women scientists are necessary and important for students—and so is a discussion about why such a high percentage of great scientists mentioned in the history books are men.

As you and your students follow modern scientific developments in newspapers, magazines, and other media, note how both women and men contribute to the advances in all fields. Make it an assumption in your class: scientists come in all genders and colors and from all countries—in short, every variety of human being. Resources abound to help you if you choose to devote a section of study exclusively to women's contributions to science.

## Generating Enthusiasm

Start each section with students' questions and ideas. What do they know? What do they want to know? Then go on and survey the history of each field you choose. The activities will emphasize science process skills and most will need little introduction—get the kids started and stay out of the way. Through the experimenting, students will be controlling variables, making predictions, recording and interpreting data, drawing conclusions, and *doing* science.

Connecting the main ideas in a historical and social context should enrich their overall understanding and make them eager to discover where science is heading today. Studying today's news should be a major goal for all of us who teach and especially those who teach science—helping students become scientifically literate and able to understand current issues and ideas.

A goal of writing and using this book is to excite the scientists of tomorrow about all there is to know now and all there is for them to discover in the future. There's a lot you can see from the shoulders of giants.

# Introduction for Students

## Science as Historical Process

What do we know and how do we know it? These two questions can lead you on a very rewarding journey. Thanks to thousands or even more years of questioning, observing, and experimenting, we know a mind-boggling amount about the world and universe around us. The average ten-year-old school child knows more science than anyone knew just a few hundred years ago. How did all that knowledge get here?

Isaac Newton (1642–1727) was born in the same year in which another famous scientist, Galileo Galilei (1564–1642), died. Responding to a question about how he could know so much, Newton is reported to have said, "If I have seen further, it is by standing on the shoulders of giants." Galileo was a giant pair of shoulders for Newton, and Newton grew giant shoulders for others himself. Every generation starts from the current knowledge and builds further.

Look outside your classroom window. The sun comes up on one side of the building, rises and travels across the sky, then heads down to set on another side. Throughout the year, the sun's path changes as it appears lower in the winter and higher in the summer. Doesn't it seem reasonable to describe the sun as traveling around the Earth?

In fact, not so long ago, most people thought the Earth was the center of the universe. Other ideas that have changed include the following:

- Scientists believe the Earth was formed about five billion years ago. In 1650, Bishop Ussher set the date at 4004 B.C.

- Things burn when they combine with *oxygen*, not because they contain a substance called *phlogiston*.

- Matter consists of tiny atoms that are themselves made of smaller substances. Earlier people believed all matter was made from four elements: earth, fire, water, and air.

- Plants make their own food mostly out of the carbon in the air, not from the soil or water.

Who knows what ideas of today will be changed in the future? Enjoy these activities and ideas that teach about how science has grown and changed, and maybe you will see something new on the shoulders of giants.

# CHAPTER 1
# Natural Selection

**TIME LINE**

| Year | Notable Event |
|---|---|
| 300s B.C. | Greeks found fossils and knew they were remains of ancient animals. |
| 1680 | The dodo became extinct. |
| 1795 | Georges Cuvier discovered bones of an ancient sea reptile. |
| 1830 | Charles Lyell published the first volume of his work claiming the Earth is millions of years old. |
| 1831 | Charles Darwin began his trip as naturalist on the HMS *Beagle*. |
| 1842 | Richard Owen used the term *dinosaur*. |
| 1859 | Charles Darwin published his book *On the Origin of Species by Means of Natural Selection*. |
| 1860 | Gregor Mendel experimented with the laws of heredity. |
| 1953 | James Watson and Francis Crick built a double helix model of DNA. |
| 1962 | Rachel Carson published *Silent Spring,* warning of environmental problems from man-made chemicals. |

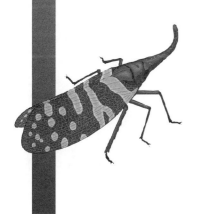

## Materials per Team

- potatoes
- toothpicks
- push pins
- pipe cleaners
- cotton balls
- other craft materials—the greater the diversity of materials, the greater the differences among the creatures

# Design a Species

## *Potato Creatures*

Ask students, "Why are specific animals successful in specific habitats?" The theory of natural selection, popularly called "the survival of the fittest," revolves around that question. If children don't suggest many survival criteria at first, guide them toward recognizing these essentials:

- ▶ ingesting and digesting food
- ▶ respiration to release the chemical energy of food
- ▶ response to stimuli
- ▶ finding a mate for reproduction to pass genes along

To begin, tell the teams that they will be designing their own creature, building it with attributes that will help it be successful. They must think about their creature: How will it survive? How will it obtain food? How will it avoid predators? What senses will it use to process information about the world? How will it attract a mate of its own species? Questions like these are central to evolution theory.

### Activity

When students are ready, provide each group with a potato and let them decide what attributes their creature will have. Make available various craft materials for them to use in creating their creatures.

When thinking about what materials to offer students, keep these two points in mind:

1. If you make many different materials available to attach to the potatoes, student groups will probably build a wide diversity of creatures. This will simulate the widely diverse set of life-forms on Earth.
2. Conversely, if you offer only a few kinds of materials, many of the potato creatures will look alike. This simulates another attribute of life on Earth. Students will recognize that many creatures are similar in overall design.

Sometimes this is a result of species diverging from common ancestors. But even very different creatures share fundamental similarities. For example, nature has provided many animals with eyes—different structures and abilities exist but there is a standard design in many species.

When students have finished their creatures, have them write descriptions of their characteristics and share them with other groups. The biodiversity of the "organisms" will be surprising. Afterward, have students bring the potato creatures outside and hide them in the environment to test their ability to blend into the habitat. (This may lead into the camouflage activity on page 11.)

Enjoy this rich activity with its opportunities for interdisciplinary extensions. To complete a cycle, students can remove materials that are not biodegradable and then bury the potatoes.

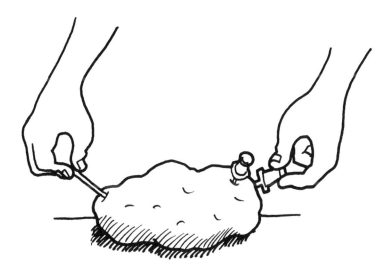

# READING:
# *Darwin's Theory*

Charles Darwin

When Charles Darwin was twenty-two years old, he began a journey around the world as the naturalist on a ship named the *Beagle*. The journey took nearly five years, and Darwin's experiences set into motion a most memorable career.

In the Galapagos Islands off the west coast of South America, Darwin studied tortoises, finches, and other animals that had lived apart from the mainland. Unique species of these animals lived on each island. Somehow the variety of conditions on different islands had produced a variety of creatures. The tortoises had different markings depending upon which island was their home. The beaks of the finches seemed fitted to their food. The Galapagos Islands continue to be a rich research area for life scientists.

Because of the animals he saw there and other experiences later in his career, Darwin formulated a theory called *natural selection.* He published the major part of it in 1859 after working on it for many years.

Darwin's book, *On the Origin of Species by Means of Natural Selection,* said that organisms live and die and many have offspring, or babies. Populations can increase very rapidly. The offspring are similar to their parents in some ways and different from them in other ways. Changes from one generation to another happen for a variety of reasons, sometimes through **mutations.**

Organisms whose changes are beneficial (helpful) to survival have a better chance to grow to adulthood and produce offspring. Organisms whose mutations reduce their ability (harmful) to survive tend to produce few or even no offspring. As generations pass, organisms whose changes aid in both survival and reproduction will dominate the species.

Darwin reasoned that variations and changes are either passed along or die out. Some people refer to this process as "survival of the fittest." Natural selection theory explains how these changes become established and passed along, and the resulting process is known as **evolution.** Evolution theory was one of the most important scientific ideas of the nineteenth century and remains a central principle in life science. Not everyone accepts the theory Darwin popularized.

Many scientists have built upon Darwin's work, interpreting life science problems and issues through natural selection theory. Darwin's contemporary, **naturalist** Alfred Russel Wallace (1823–1913), explored the Amazon River area and developed a theory of evolution at about the same time as Darwin. At first, there was much resistance to the idea that species change over time. More than a century after the publication of his famous book, Charles Darwin's theories continue to be disputed by some educators and scientists.

Darwin read the work of Charles Lyell (1797–1875), a geologist of his time who argued that the Earth was very old. This idea was not easily accepted by the public, either. Evolution usually depends upon very great stretches of time to do its work. Over time, the competition in nature selects the organisms best suited to succeed. Not only individual organisms but entire species may eventually fail to produce successful offspring. When that happens, the species becomes extinct. Darwin worked with fossils of extinct creatures, and that experience also helped him see changes over time.

Let's apply the theory of evolution to an animal with a distinctive set of characteristics. Take the case of giraffes. They are born with

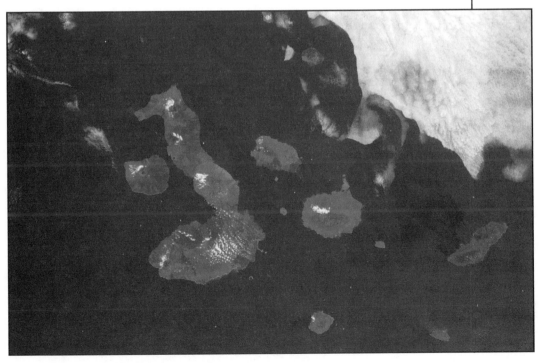

Galapagos Islands (Courtesy of NASA)

long necks, well adapted for eating leaves high in the trees on the African plains. A short giraffe would not compete for food successfully against its taller neighbors and could not easily raise a successful family. Its neck would not grow much by stretching up for leaves, as some of Darwin's opponents claimed. Natural selection would favor the taller giraffes. (Male giraffes also use their long necks to fight one another, so a long-necked male is favored as a fighter.)

Can you think of some animals for whom a long neck would be a disadvantage? All animals have qualities that allow them to find food, defend or protect themselves, and meet others of their species in order to pass along their genes to offspring. Start with a potato and model a creature you think is well adapted for life in a habitat you create.

## Vocabulary Words

evolution ................................. the theory that genetic changes from generation to generation over time cause species to change gradually

mutation ................................. permanent change in the genes of an organism

naturalist ................................. scientist who studies natural objects and organisms

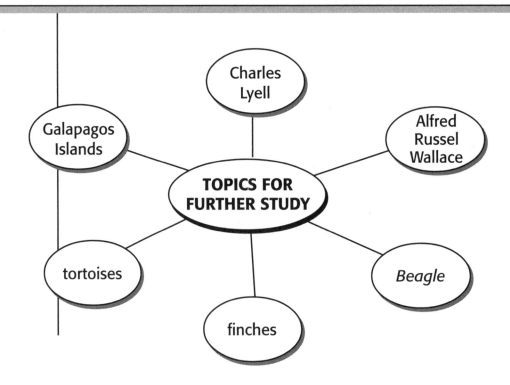

# Camouflage

## *Blending In*

Many prey species have evolved coloring that camouflages them from predators. Some predators use their ability to disappear into the background as a way to surprise unsuspecting victims. This simulation can generate data enabling students to track a rapid case of natural selection.

### Materials per Team

- set of colored disks or paper circles for each group
- paper or fabric to use as background
- paper cups

---

### Activity

Have students use hole punches or scissors to make sets of colored disks. Start each group with ten disks of each of ten different colors. For example, use red, yellow, orange, green, blue, purple, black, white, brown, and gray. Keep extra disks sorted in cups for use in future rounds. Have each team mix their hundred disks in an empty cup and then pour them out onto a background of paper or fabric, scattering the disks randomly. Members of the group then take turns removing the ten easiest to find disks. After students remove fifty disks, round one is over.

11

Have students sort the remaining disks by color. For every pair of similarly colored disks remaining, students can add two new disks of that color. This simulates the reproductive success of animals that can escape predation. Count and record the number of disks. Single disks do not reproduce but match enough of them to bring the population back to one hundred for round two. This can simulate creatures moving into the area and also higher numbers of offspring for some survivors. Put the new population into the mixing cup to prepare for round two.

Make a chart or graph of the population at the beginning of each round and the pattern will become clear. Because the population of each round will be one hundred, the percentage of each color will be simply the number of disks.

| Population Chart | | | | |
|---|---|---|---|---|
| **Color** | **No. at Start** | **Starting Round 2** | **Starting Round 3** | **Starting Round 4** |
| red | 10 | | | |
| orange | 10 | | | |
| yellow | 10 | | | |
| green | 10 | | | |
| blue | 10 | | | |
| purple | 10 | | | |
| black | 10 | | | |
| white | 10 | | | |
| tan | 10 | | | |
| gray | 10 | | | |

After a few rounds, the population will be dominated by the colored disks that are hardest to find against their background. The most effectively camouflaged disks will have been naturally selected to succeed in their environment.

## READING:
# Camouflage

## Variation, Heredity, and Survival

Two central themes in evolution theory are variation and heredity.

### Variation

Charles Darwin and other nineteenth-century naturalists could see that organisms were not all alike. When most animals had a group of babies, the babies differed from one another. Seeds from a plant do not produce exact copies, either.

Darwin theorized that trying to survive in nature tested these variations. Sometimes a variation helped and sometimes it didn't.

### Heredity

Although offspring vary, they do resemble their parents in many ways. Darwin did not know the way characteristics are passed on to the next generation. But he did know that farmers and pet owners could breed their animals and "improve the stock." It was common in Darwin's nineteenth-century England to pair up the best animals to ensure that the next generation would have the traits owners wanted.

Heredity is the process by which traits are passed along from one generation to another. Successful variations are preserved in a species by being carried in individuals. Darwin did not know about **genes** and **DNA,** but he developed a powerful theory with the concepts of variation and heredity.

In the Galapagos Islands, the process of evolution can work quickly. Species there are isolated and develop special niches on islands. The finches on different islands have been found to change rapidly. They change as a species from generation to generation, responding to climatic conditions. Those finches whose traits allow them to survive the rigors of life on their island produce offspring with a chance of survival.

Rapid evolution has occurred in other parts of the world also. Certain species of moths rely on camouflage to avoid being detected and eaten by predators. Birds eat the most noticeable moths. The moths good at hiding, those that blend in best to their surroundings, have a better chance of surviving to reproduce.

British naturalists in the nineteenth century noted that the population of a common moth called the salt and pepper moth, or peppered moth, was changing. More black individuals were appearing near cities, replacing some of the speckled, bi-colored ones.

Evolution theory suggests that the black moths began to be "selected" to reproduce because their black color became more useful. In the 1800s, British cities produced lots of soot from the coal fires powering factories. The soot settled onto trees, blackening their trunks and providing hiding places for the black moths.

In the 1950s, biologist H. B. D. Kettlewell (1907–1979) did a study supporting this evolutionary behavior. Some studies were later questioned as to whether the scientists reported true results, but clearly species have changed in response to changes in their habitat. Darwin's natural selection theory explains that successful creatures survive and reproduce. In that way, conditions in nature select the way species change.

Scientists often study species that have been isolated for a long time to see how they have changed. Two interesting examples are the cichlids (pronounced sick-lids), which are perchlike fish that live in Lake Victoria in Africa, and honey creepers, birds of Hawaii.

When you study interesting species, consider how the differences among individuals might be either helpful or harmful to their survival.

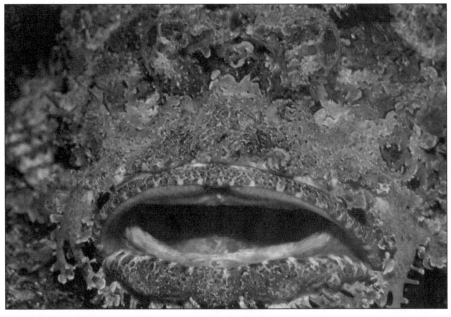

While the anglerfish attracts prey with dangling lures attached to its body, its powerful jaws blend into its surroundings.

## Vocabulary Words

DNA ........................................... deoxyribonucleic acid; a double twisted helix inside a cell that contains the genetic code for that organism

genes ........................................ sections of DNA that produce a trait in an organism

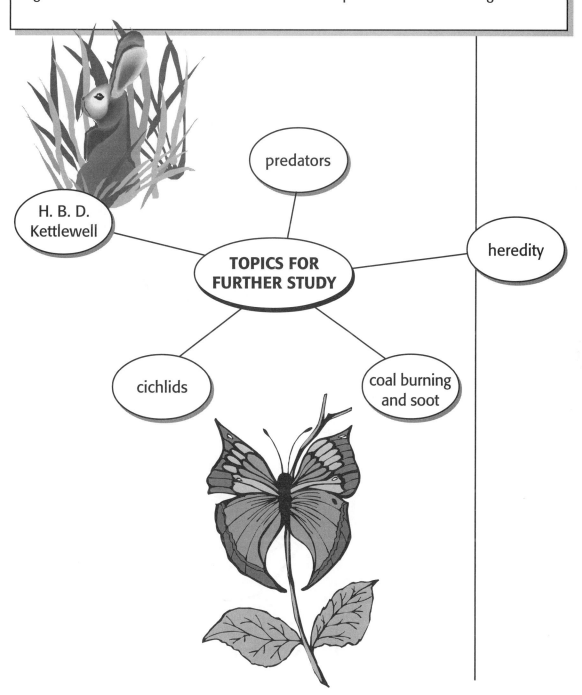

predators

H. B. D. Kettlewell

**TOPICS FOR FURTHER STUDY**

heredity

cichlids

coal burning and soot

From *Science Giants: Life Science* © Good Year Books. This page may be reproduced for classroom use only by the actual purchaser of the book. www.goodyearbooks.com

## Materials per Team

- photos of animals from old nature magazines or calendars
- construction paper
- scissors
- glue sticks

# Adaptation

## *Sneak a Peek*

All organisms are special. The forces of natural selection have ensured that each species is well designed to survive in its habitat. Here's a way for students to concentrate on the unique characteristics of various creatures.

## Activity

Ask students to collect pictures of animals from nature magazines and other sources. Zoo brochures, calendars, advertisements, and travel literature are other likely places to find high-quality photos. Encourage students to study the pictures for unusual and interesting attributes of each species. Birds have evolved beaks adapted to specific food sources. Predators have strong jaws or sharp claws and talons. Animals in snowy climates may have broad, furry paws. Camouflage strategies are common in the animal kingdom.

Have students each cut out an animal. Then they should fold a piece of construction paper like a book so that the photograph can be glued or stapled inside and covered when the other half is folded over. Challenge students to carefully cut a hole from the top layer so that they create a window revealing only a specially adapted part of the animal inside. Students then show each other their "Sneak a Peek" folders, guessing what the animal inside will be.

Have students trade folders in this class and with other classes. You will have an entertaining library illustrating how evolution has populated the world with an amazing variety of life.

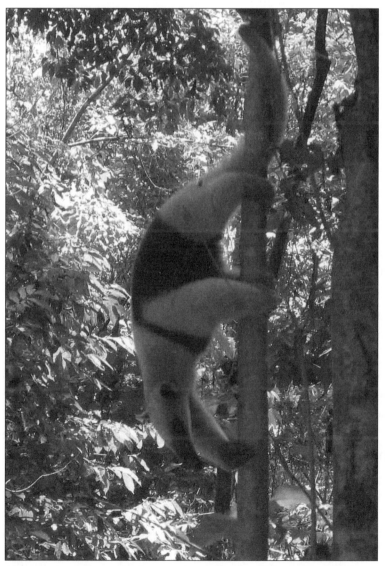

A Tamandua anteater in Corcovado National Park in Costa Rica
(Dirk cander Made, courtesy of Wikipedia)

*READING:*

# The Diversity of Life

How can there be such diversity of life on our planet? Every year people discover new species. Every year we learn more about the amazing traits of species already identified.

Today, most scientists believe that the Earth is very old. It has been a great challenge trying to determine the age of our planet and calculating when life first appeared. **Fossils** recovered through the early 1990s indicate that simple organisms like bacteria may have been present 3.5 billion (3,500,000,000) years ago. Given that much time, a lot can happen.

Charles Darwin reasoned that over such a vast amount of time, countless types of life-forms could have arisen. He based his theory of **evolution** upon three assumptions:

1. **Variation:** Members of a species vary, or differ, from one another. Individuals are not all alike.
2. **Heredity:** Parents pass along traits to their offspring. Characteristics, or attributes, of a parent can be inherited by its descendants.
3. **Different Number of Offspring:** Some of the members of a species will have more offspring and some will have fewer.

During the process of reproduction, changes (called **mutations**) of cells may occur. Mutations can be caused by outside factors (environment, chemicals, radiation) or by errors in the copying process within a cell. If the change helps the organism, it may become incorporated into the species. The carrier of the mutation will reproduce and pass on the new improved characteristic.

What will the future inhabitants of the Earth look like? What mutations and changes will occur and become part of the code of a species life? Fossils give us clues about amazing animals of the past. You can research many unusual creatures that are alive today. But no one knows what the animals of the future will be.

## Vocabulary Words

evolution ................................... the theory that genetic changes from generation to generation over time cause species to change gradually

fossil ........................................ the remains of ancient organisms

mutation ................................... permanent change in the genes of an organism

heredity ................................... process by which traits are passed from one generation to another through genetic information

variation ................................... difference or change in form or structure from usual type

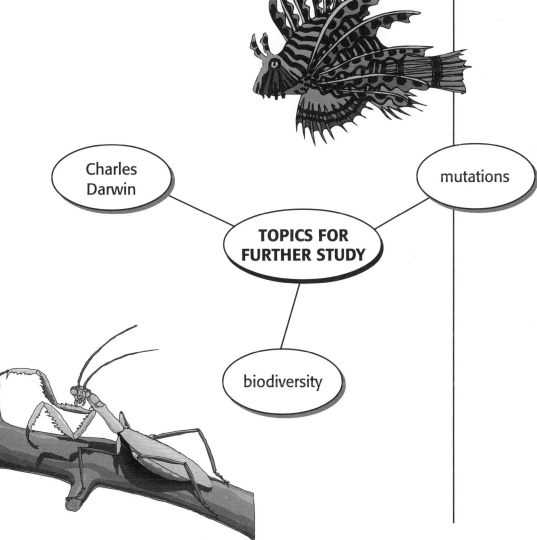

Charles Darwin

mutations

**TOPICS FOR FURTHER STUDY**

biodiversity

## Materials per Team

- pumpkins
- paper cups
- scale
- measuring tapes or meter sticks

# Individual Differences

## *A Tale of Two Pumpkins*

This is a good activity for October but it will make a strong impression on the students at any time of year. Choose at least two pumpkins of contrasting sizes to compare. Set up a data chart so students can compare pumpkins. Students may suggest ways they can compare the pumpkins.

### Activity

Include the following activities in addition to those the students suggest. Have them find the mass of each pumpkin using a scale or balance. Then they should measure the circumference and height (not counting the stem). Then they can scoop out the seeds and pulp of each one and find the resulting mass, called the *inside mass.* Finally, have them count the seeds in each pumpkin and record the result.

|  | Mass | Mass When Hollowed | Circumference | Height | No. of Seeds |
|---|---|---|---|---|---|
| Pumpkin A |  |  |  |  |  |
| Pumpkin B |  |  |  |  |  |
| Pumpkin C |  |  |  |  |  |
| Pumpkin D |  |  |  |  |  |

Students can use their calculators to discover many interesting facts. What is the ratio between some of the measures? For example, is the inside mass usually a fixed proportion of the total mass? Do larger pumpkins tend to have more seeds than smaller ones? Generating hypotheses about the data and then testing them creates a rich math-science connection derived from a real-life activity.

Most pumpkins produce hundreds of seeds. A pumpkin plant, as all other organisms do, must provide a way for its genes to be passed along to another generation. Producing an abundant number of seeds raises the odds that at least one will germinate (sprout) and grow to produce its own offspring. Students can set up experiments with the seeds after they harvest them, but hybrid varieties of pumpkin may not sprout.

## READING:
# *Building on Darwin's Theory*

Since Charles Darwin's time, we have learned a lot about how evolution proceeds. Darwin knew parents pass on traits to their offspring but he didn't know how. He would be amazed at all we've learned.

All living things are made of cells. In the twentieth century, scientists learned that parts of cells called **chromosomes** contain **genes,** which are made of **molecules** called **DNA.** DNA molecules can be copied within cells so that coded information is passed along when the cells reproduce. When two parents produce offspring, some characteristics of each parent are passed along to the next generation, coded on the DNA.

Most pumpkins produce hundreds of seeds. A pumpkin plant, as all other organisms do, must provide for its genes to be passed along to another generation. Producing an abundant number of seeds, all carrying the plant's DNA, raises the odds that at least one will **germinate** and grow to produce its own offspring.

Darwin called the process by which variations and changes are either passed along or die out **natural selection.** Some people refer to this process as "survival of the fittest." With pumpkins, a good strong plant will mature, flower, and grow seeds. More seeds produced generally means more chances to reproduce. A weak plant might fall prey to disease, **drought,** or other misfortune.

You now have a good idea of how the natural selection theory explains how changes become established and passed along. The theory rests on Darwin's three basic assumptions.

**Variation**: differences among individuals

**Heredity**: traits can be inherited

**Different Number of Offspring**: success varies

Organisms whose changes are beneficial tend to grow to adulthood and produce offspring. Organisms whose **mutations** reduce their ability to survive tend to produce few or even no offspring. As generations pass, organisms whose changes aid in survival and reproduction dominate the species.

The theory of evolution has been around for well over a hundred years, but not everyone believes it. Many life scientists do, however, and the theory has been able to explain a great deal about the history and direction of life on Earth. But **controversy** continues.

Edward Wilson (b. 1929), an American biologist who studied ants extensively, has theorized that much human behavior is guided by evolution. Wilson called this field of study **sociobiology.** Wilson's work continued the tradition of riling up opposition, just as Darwin's theories did.

Stephen Jay Gould (1941–2002) and Niles Eldredge (b. 1943) proposed a theory that evolution moves in bursts rather than gradual steps, a process called "punctuated equilibrium." Instead of evolution being a long, slow, steady process of species changing, Gould and Eldredge thought periods of rapid change are separated by long ages of **stability.**

What will be the next great changes in our thinking about how life changes on Earth? Will humans find evidence of life elsewhere in the universe and apply evolution theory to another planet?

## Vocabulary Words

| | |
|---|---|
| chromosomes | structures in cells that contain genetic information |
| controversy | dispute; difference of opinion |
| DNA | deoxyribonucleic acid; a double twisted helix inside a cell that contains the genetic code for that organism |
| drought | prolonged period without rain |
| genes | sections of DNA that produce a trait in an organism |
| germinate | sprout |
| heredity | process by which traits are passed from one generation to another through genetic information |
| molecule | two or more atoms bonded together |
| mutation | permanent change in the genes of an organism |
| natural selection | the survival of individuals whose characteristics are advantageous for their environment and elimination of those individuals who do not succeed |
| sociobiology | the analysis of human behavior from the viewpoint of evolution theory |
| stability | steadiness; resistance to change |
| variation | difference or change in form or structure from unusual type |

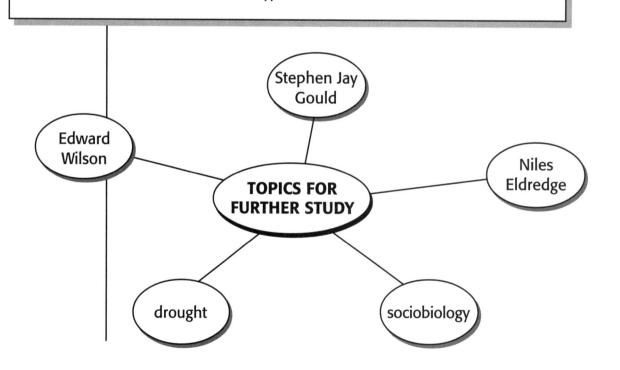

# CHAPTER 2
# Classification

**TIME LINE**

| Year | Notable Event |
|------|---------------|
| 300s B.C. | Aristotle classified animals according to their characteristics. |
| C. A.D. 55 | Pliny the Elder wrote a large science text that included animals that were both real and mythical. |
| 1667 | John Ray published a classification book dividing things into animal, vegetable, and mineral categories. |
| 1680 | The dodo became extinct. |
| 1735 | Carolus Linnaeus published his classification system. |
| 1795 | Georges Cuvier developed a method for classifying mammals. |
| 1811 | Eleven-year-old Mary Anning discovered icthyosaur bones. |
| 1859 | Charles Darwin published *On the Origin of Species by Means of Natural Selection*. |
| 1866 | Ernst Haeckel hypothesized that embryos undergo a process that mirrors evolution. |
| 1956 | Herbert Copeland suggested that two kingdoms of microbes, bacteria and protists, be considered when classifying organisms. |
| 1959 | Robert Whittaker placed fungi in their own kingdom in his classification scheme. |

## Materials per Team

- paper and pencils

# Classifying Children

## *You Are Unique*

Everyone is known by a name. The students in a classroom are all unique and identified specifically by their first name and family name. In this activity you can help them understand hierarchical classification by grouping them in increasingly broader (or narrower) categories.

## Activity

Starting from either end of the list, have students generate data about themselves in the following categories:

first name • family name • classroom • school • town state, province, or territory • country • continent • planet

Ask different students to give their answers in specific categories. From "planet" to "classroom," everybody may be the same. But most likely everybody has a different name. Ask students to generate other classification categories, such as "street" or perhaps "place of birth" for the geographical categories.

Classifying students this way rather than by physical characteristics reinforces that all human beings are members of one species. Variations are almost limitless: color of skin, eyes, and hair; height and weight; sound of voice; arrangement of teeth; the list goes on and on. Plants of the same species grow different-color flowers, pets in the same litter have different coloring, and humans follow this fact of life. Organisms within the same species vary. Viva la difference!

# READING:
# Scientific Names

Imagine that the students in your school had no names. People might refer to "that tall girl in the fifth grade," or "the boy with red hair who always brings his lunch." What a mess!

Naming plants and animals can be like that too. Sometimes two or more different animals have the same name. A wildcat in one part of the world might be a different species from a wildcat somewhere else. Sometimes the same animal has more than one name. For example, *woodchuck* and *groundhog* are two names for the same familiar American rodent.

Life scientists around the world do have a system to identify species with a name everyone agrees upon. This system is called scientific **classification,** and it began in the 1750s.

Carolus Linnaeus (1707–1778) was a naturalist who collected and studied many specimens. He wrote books about plants and animals and assigned each different living thing a two-part name. The names Linnaeus chose were Latin, a language that was not popularly spoken but was widely used in churches, schools, and official writing. He even changed his name from its original spelling Carl Linné to a Latin version.

The designation of species is the critical unit of classification. In general, groups of animals or plants that breed together and produce offspring together in the wild belong to the same species. Different breeds of dogs, for example, can mate and have puppies, and the same is true for cats producing kittens. The offspring will inherit a mixture of the mother's and father's characteristics. But a dog and a cat cannot produce a set of babies together that would be part dog and part cat.

Although many of Linnaeus's groupings have been changed, his basic system of classification remains standard worldwide. Scientists everywhere know which organism another scientist is referring to simply by knowing the two-part Latin name.

## Vocabulary Word

classification ...........................system of grouping organisms based on their similar characteristics

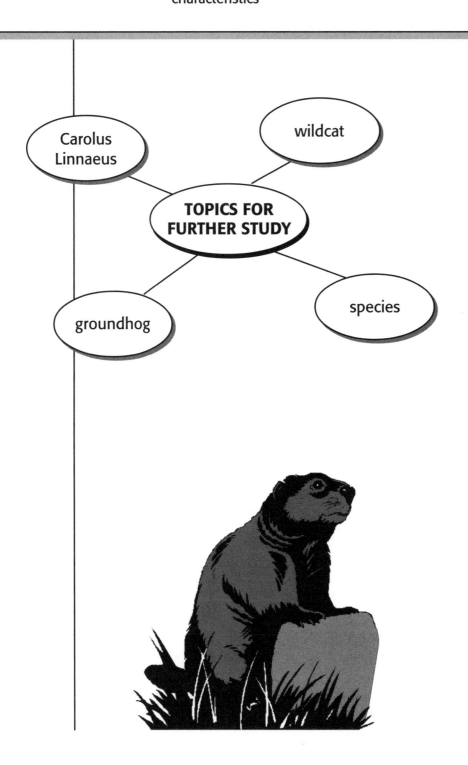

Carolus Linnaeus

wildcat

**TOPICS FOR FURTHER STUDY**

groundhog

species

# Categories

## *Sorting the Species (Animals)*

Historically, scientists have built classification systems primarily by grouping organisms according to common characteristics. Larger categories share many traits, and the process narrows down until a unique species is identified and named. In this activity, students will sort animals into groups and research the scientific name for the phylum and class of each group.

### Activity

Assemble a large selection of pictures of animals. Commercial sets of cards are available and some publishers have reproducible drawings you can copy. Computer software often contains files of animal pictures also. If you gather the pictures yourself, save them for the "Sneak a Peek" activity on page 16.

Distribute a random set to each team of students. Instruct them to group their pictures according to attributes about the animals. Encourage discussion and share ideas among the class. Explain to the class that you would like them to find out how scientists classify animals into categories.

### Materials per Team

- photos or drawings of animals (nature magazines are a good source)
- encyclopedias or other reference tools

In the Linnaean classification system, the first category is *kingdom*, followed by *phylum* (plural, *phyla*). Using encyclopedias or other research tools, have students look for the scientific names and common descriptions of phyla in the animal kingdom. They should then write the names on cards and sort their pictures into the proper phyla. They can use the encyclopedias to look up the classification of questionable species. Many of the animals will fall within the phylum *Chordata*, which contains animals with backbones and are commonly called *vertebrates*.

Help students further classify their vertebrates (animals with backbones) into *classes* through some further fact finding. Common classes of vertebrates are *b*irds, *f*ish, *a*mphibians, *r*eptiles, and *m*ammals. The acronym B-FARM will help students remember the Chordata classes.

Share with students the names of the remaining categories in this classification system: order, family, genus, and species. As a conclusion, share with students how human beings are classified:

Kingdom ...... animalia
Phylum ......... chordata
Class ............. mammalia
Order ............ primates
Family ........... hominids
Genus ........... homo
Species ......... sapiens

Students can make posters and mobiles to display in the classroom and show the classification groups they've learned.

### READING:
# Grouping Living Things

Sorting things is a natural activity for humans. It helps us understand the world and see how different things are connected. To sort something as complicated as the set of all living things, people have tried lots of schemes.

Perhaps the first sorting activity was to divide possible food sources into edible versus harmful. If you had to gather or hunt your own food, imagine how important it would be to identify species correctly. When you were bitten or scratched, you would need to know if the wound was poisonous. What is the **antidote**? So, by necessity, humans have been naming and classifying their fellow organisms for a long time.

Aristotle (384–322 B.C.E.) tried his hand at classification back in the 300s B.C. He and his fellow scientists knew of about one thousand species of plants and animals. Aristotle divided animals into two categories—those with backbones and red blood and those without. He arranged plants by size. Pliny the Elder of Rome (A.D. 23–79) wrote a thirty-seven-volume text about animals, some of which were completely **mythical.**

In the 1600s, John Ray (1628–1705) of England began to define species. Ray divided plants into categories based on whether their seeds contained one leaf or two. But the main architect of the modern scientific classification system was Carolus Linnaeus (1707–1778).

Linnaeus's great contribution to science was the **binomial** system of classification. He began to assign all organisms a two-part name consisting of **genus** and **species.** Linnaeus used Latin, a language taught widely in Europe for many years, rather than local languages. Scientists were able to identify organisms and have people in other countries use the same scientific name.

When you sort things, do you start from major characteristics or do you hunt for little things? It seems obvious to start with the most noticeable things. That's been the strategy throughout history. But things are changing.

When Linnaeus, Aristotle, Ray, and most other scientists before 1850 grouped organisms, they probably didn't know about evolution. Now, when scientists classify a species, one of the first things they consider is what common ancestors it shares with other species. Using a microscope has caused other major changes in classification. Looking at the **cells** of an animal or plant can reveal similarities and differences undetected by the eye alone.

The roster of species in the world does not stay the same. New species are discovered every year. Unfortunately, many species are endangered and some have become extinct. When scientists uncover fossils of extinct species, they try to fit them into the classification scheme somewhere. Evolution theory attempts to explain how modern species have descended from those in ancient, pre-human times.

Humans have experimented with **hybrids** for centuries. Hybrids are organisms from different species deliberately interbred to create offspring of mixed characteristics. In the wild, one way new species might be created occurs when natural barriers separate populations of animals. The isolated groups evolve differently from the original species. Islands tend to have unusual organisms because the water around them can keep some foreign species from establishing themselves there.

Keep your eyes and ears open for news about discoveries of species and how they are classified. You might think this branch of science is pretty **static** but ideas are changing rapidly. Can you imagine birds as the relatives of dinosaurs? Is a panda bear really a bear? How long have horseshoe crabs roamed the oceans? There are many interesting questions that the classification system helps us investigate.

# Vocabulary Words

antidote ........................................ something that stops the action of a poison

binomial ...................................... having two names. In taxonomy, a two-part name consists of genus and species.

cell................................................ the smallest, microscopic-sized unit of organisms

genus ........................................... category of organisms into which related species are grouped. Each type of organism has a scientific name consisting of its genus and species.

hybrid .......................................... offspring produced by parents who are genetically different

mythical...................................... from a story, usually about a hero or supernatural creature

species ........................................ group of organisms that is considered one type. Generally, organisms within a species can breed among themselves.

static ........................................... unchanging

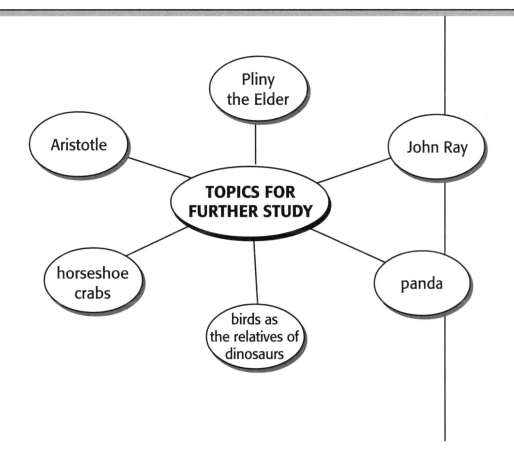

**Materials per Team**

- bird identification books, photos, and posters
- paper and pencils

# Bird Beaks

## *Examine the Bill Carefully*

Evolution theory is based on a utilitarian view of animal and plant structure. The phrase "form follows function" usually guides scientists seeking clues to understanding behavior. When biologists study the connection between the bodies and behavior of animals, they often gain insight about evolution and classification. Giraffes have long necks and can access food sources high off the ground. Zebras are shorter and graze at ground level.

Birds have evolved wings to fly and need to perch on their feet. That means their beaks are especially vital to food gathering. The shape and size of a bird's beak reveals much about what and how it eats. In this activity, students can practice drawing bird beaks and matching them to their specific uses (see page 35).

## Activity

To create model sketches on a chalkboard or overhead projector, draw or photocopy the examples provided here. Older students may be able to use resource materials, such as bird identification books to make their drawings independently. For a further extension of the activity, have students research and sketch bird feet, matching the forms to their functions.

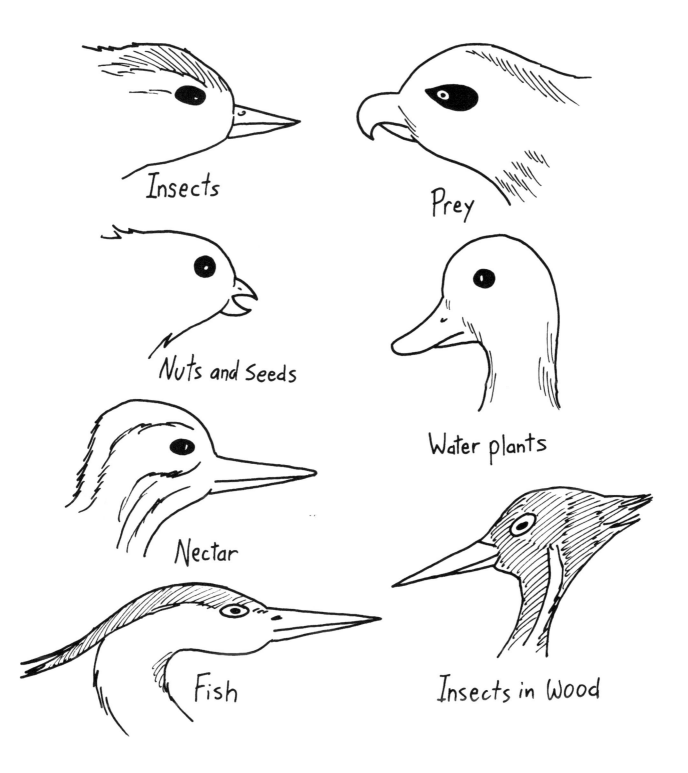

Insects

Prey

Nuts and Seeds

Water plants

Nectar

Fish

Insects in Wood

*READING:*
# The Right Equipment to Survive

When Charles Darwin explored the Galapagos Islands, he noticed that the beaks of finches and mockingbirds differed from island to island. The governor of the region told Darwin that the giant tortoises on each island had shell designs that were unique to that particular island. Separated by deep water, the islands are **volcanic** peaks that reach up out of the Pacific Ocean. Because of their separation, many of the creatures inhabiting the Galapagos have evolved into different species.

Returning to England, Darwin continued thinking about how new species develop, a process called *divergence.* Studying pigeons, barnacles, and other animals, he reasoned that the uniqueness of each species was the result of natural selection. Characteristics that helped the successful members of the species to survive were passed on to offspring. In the case of the Galapagos birds, their beaks enabled them to eat the particular food that grew on their particular island.

Darwin's famous theory states that through the process of evolution, animals have adapted to their food sources with very specific body parts and behaviors. This idea revolutionized life science study, and scientists have argued about it ever since it was published.

Twentieth-century field biologists studied the Galapagos finches carefully. They discovered that the harsh conditions place intense selection pressure on the birds. *Selection pressure* means that severe conditions limit the number of survivors. Only the birds best equipped to live will make it to full adulthood and produce offspring during hard times.

On the Galapagos Islands, drought and floods have sometimes limited the production of particular seeds, the food source for the birds. A bird has a better chance of surviving difficult times if its beak can open a variety of seeds. Researchers found that small differences in beak size could make a big difference in a bird's ability to eat seeds.

Natural selection pressure works quickly in the Galapagos. Beak size among members of some species changed in a few years. Those birds whose beaks enabled them to eat the seeds that grew during the drought survived and reproduced. Others died out.

Selection pressure may also increase the birth of **hybrids,** or the interbreeding of different species. During hard times, birds may be forced to seek food in different ranges. When that happens, individuals from different species may enter each other's territory and pair up.

Studying classification systems and evolution theory helps us understand how organisms are related and how they fit into their environments. In the middle 1800s, Darwin and others took a leap to the conclusion that animals and plants are the way they are because of natural selection. They succeed in growing to adulthood and can reproduce a new generation. Unsuccessful organisms do not succeed and reproduce. The Galapagos Islands are a natural laboratory for studying this process because of their isolation. Africa's Lake Victoria and its cichlid fish provide another example of evolution in action. Can you find other examples of isolated species and how they have changed? (Key search words might be **adaptation, evolution,** and **natural selection.**)

Find out about the birds in your area. Study their beaks and how they eat. Watch them at a bird feeder. Soon you may be able to tell what a bird eats by the shape and size of its beak.

## Vocabulary Words

adaptation ............................. a trait of a species that helps it survive; behavioral change

evolution ................................. the theory that genetic changes from generation to generation over time cause species to change gradually

hybrid ...................................... offspring produced by parents who are genetically different

natural selection .................... the survival of individuals whose characteristics are advantageous for their environment and elimination of those individuals who do not succeed

volcanic ................................... from volcanoes, or from rock formed by volcanoes

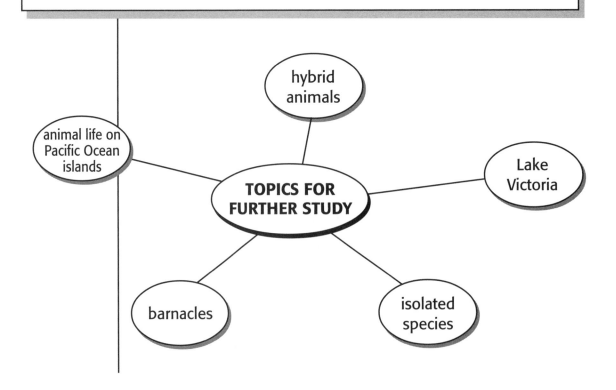

# Trees

## *Sorting the Species (Trees)*

This activity gives students some hands-on experience with the skills necessary for classification. They will need to examine their branches carefully to determine which attributes are similar and which are different.

## Activity

Tell students that you have trimmed some trees to study. The cutting should help the trees grow in a healthy, attractive way. To learn more about different kinds of trees, ask students to sort the branches you have cut. Distribute branches randomly to students and ask them to find other students who have the same kind of branch.

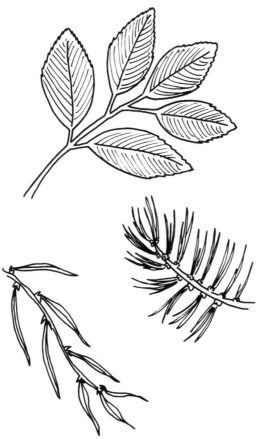

### Materials per Team

- selection of tree branches, three to six samples of each tree species, enough for one for each student (Sort these before the activity.)
- tree identification guides

Once the groups with similar specimens have formed, instruct each student to talk with teammates and make a detailed description and sketch of his or her branch. Each group should compile a written list of attributes to be reproduced or written on chart paper.

Next, assemble groups consisting of students with different branch samples. (You can do this during a later class period, depending on scheduling constraints.) Have the new groups work cooperatively to compile lists of attributes that are common to some of the branches but not others. Can they find some descriptions that only apply to one of the species? Senses of smell and touch help, too. Ask: How does the bark feel? Can you close your eyes and identify the tree type by smell?

Students can then try out their skills by hunting for their species in the wild. Display published tree guides and they'll see how the naturalists who write them use classification skills to teach species identification.

Use your judgement about how many different species to use. Younger students may learn a lot from only two or three species of quite varied type. Older students can be challenged by species that are similar, and they can try to create their own identification key.

Students often find identification keys helpful with tree branches. These guides ask a series of questions about the specimen. Each answer leads to either another question or the name of the tree.

## READING:
# How Living Things Are Sorted

Classifying is a tricky skill to master. To go from broad categories (like trees) to medium categories (like **coniferous** trees) and finally to species (white pine) can be challenging. It's worth it to be precise though, especially to scientists. You will appreciate the diversity among Earth's creatures if you study how they are categorized. **Taxonomy** is the name for the science of classification.

The levels of sorting for the animal kingdom, known as the kingdom *Animalia*, are phylum, class, order, family, genus, and species. Can you think up a **mnemonic** device to remember the names? One famous sentence reads "*King Phillip Came Over For Ginger Snaps.*" In four other kingdoms—plants, **protists, fungi,** and **monerans**—the word *division* is used instead of phylum.

What do each of these categories mean? Are the words *protist* and *moneran* unfamiliar and strange? For many years, scientists used a system with just two kingdoms, animals and plants. More common now is a five-kingdom scheme, adding fungi, protists, and monerans to the system because many organisms didn't seem to fit in the original two kingdoms. Some scientists favor a three-part classification system based on cell structure.

Kingdom is the largest grouping. After that, the grouping division, or phylum, sorts organisms by a set of major characteristics. Then the sorting continues down through each level of grouping. Although there can be subspecies, the two most exclusive units are **genus,** or group, and **species,** or kind. Genus and species combine to form the scientific name of the organism.

Choose an animal or a plant to research. Look up information about it and you can trace its classification from species level up to kingdom, or from kingdom down. You should be able to learn the characteristics that make it eligible for membership in each level. Our species, *Homo sapiens,* shares characteristics with a whole kingdom of animals, but by the time you travel through the categories, we are a unique species.

## Vocabulary Words

coniferous ................................. referring to a cone-bearing tree

fungus (pl., *fungi*) ................. organism from the kingdom Fungi. This kingdom includes mushrooms, yeasts, molds, and other organisms.

genus .......................................... category of organisms into which related species are grouped. Each type of organism has a scientific name consisting of its genus and species.

mnemonics ............................. devices, symbols, reminders, and so on, that can be used to help remember something

monerans ................................. bacteria, organisms in the kingdom Monera

protists ...................................... organisms from the kingdom Protista, a group that includes mostly one-celled organisms with a nucleus

species ....................................... group of organisms that is considered one type. Generally, organisms within a species can breed among themselves.

taxonomy ................................. science of classifying organisms

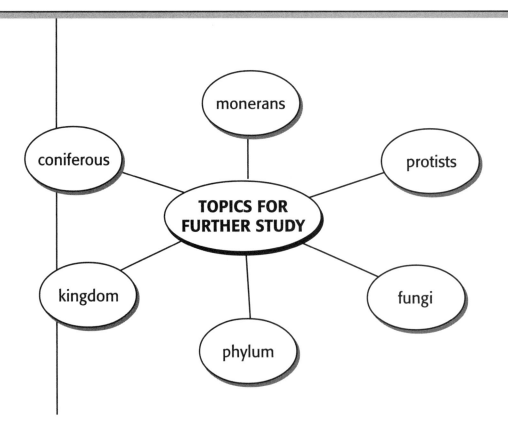

# Five Kingdoms

## *Classifying an Organism*

Students learn classification skills very early. Some of the first structured activities in preschool involve sorting and matching. The scientific discipline of classification can be introduced at almost any grade level because it builds on those early skills. Scientists arrange organisms into ever-narrowing sets. Beginning with the broadest category, kingdom, similar organisms are grouped until a unique species is identified.

The groupings are based on evolutionary development. The more closely related species are, the farther down the hierarchy they remain grouped.

### Materials per Team

- index cards
- access to photos or clip art
- scissors
- research tools (encyclopedia, Internet, library, etc.)

## Activity

In this activity, students glue animal or plant pictures to one side of a card and write the organism's taxonomy on the other. You can decide how much involvement students will have in creating the materials. You can copy a master sheet and have students cut and glue the pictures to index cards. Older students can find or draw their own pictures. Assign students to write classification information in either case.

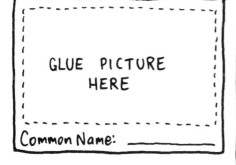

GLUE PICTURE HERE

Common Name: _____

STUDENT NAME _____
KINGDOM _____
PHYLUM or DIVISION _____
CLASS _____
ORDER _____
FAMILY _____
GENUS _____
SPECIES _____

You may have teams work on this project. Individual students can choose related but different animals from their teammates and discover how the differences affect classification. They can search CD-ROM disks, Internet sites, magazines, encyclopedias, and books for photos or drawings, and most encyclopedias include a classification chart. Legally copied or downloaded clip art can be resized on the computer to fit the index cards, a useful technology skill for students to practice as they learn life science concepts.

Which names to use and teach in taxonomy presents another choice. Linnaeus used Latin in his scheme—it was the closest thing to a universal scientific language in the 1700s. But it will be more meaningful for beginning scientists to use familiar names and concentrate on the relationships, not the nomenclature. With older students, you can include the Latin names for classification categories and explore the etymology of scientific terms.

The Latin names represent something important or descriptive about the organism. Some names include major characteristics. In other cases, a name commemorates a geographical location or person associated with the discovery and study of the animal or plant.

*READING:*
# Understanding Classification

*How many kinds of animals and plants are there?*
*What is the biggest animal?*
*What are the smallest organisms?*

Have you ever wondered about these questions? The diversity of life on Earth is enormous. Many creatures seem as if they could be part of a science fiction movie. Scientists discover new species every year, and each new species is placed into a sort of filing system. Classifying species is very important for lots of reasons—for example, trying to answer those questions we just asked. How did the current system come about, and how does it work?

Organisms have different names in different languages. Even within one country, people in one locale may refer to a species by a name unfamiliar to people living elsewhere. But scientists all over the world use an identification system that names each species precisely. In addition, this classification system places organisms in a kind of family tree. Animals or plants that descended from common ancestors are placed in categories close together. Scientists try to learn how newly discovered species are related to known species.

Credit for devising this naming system is given to Carl Linné (1707–1778), a Swedish scientist from the eighteenth century who is known by the Latin name Carolus Linnaeus. Linnaeus assigned each species a **binomial** name: genus (group) and species (kind).

Once a name is established for a species, it joins other species in larger groupings. This process continues with each level containing a broader set of characteristics. Kingdoms are the largest groups. Since the later part of the twentieth century, scientists often classify organisms into five kingdoms. Before we set the **hierarchy**, let's think of an analogy to understand the classification system better.

Imagine a campus of five separate buildings, sort of like a small college. This campus stores a page of information about every species discovered so far. The buildings represent the places where the five major types of organisms are classified. The buildings are labeled *animals, plants, fungi, protists,* and *monerans.* The protists and monerans are simple, one-celled creatures. There are differences between these two kingdoms, especially in the nucleus of their cells.

Perhaps in the far distant past, there was only one building with one kind of creature. But life has become more varied and complex, and now there are five buildings. Each has records for one *kingdom.*

Inside each building are separate rooms. Each room houses a different *phylum.* Organisms in each room are quite different. In each room, there are filing cabinets. Each filing cabinet contains records of a different *class.* Some filing cabinets are tall and others are short but they all have drawers that pull out. Each drawer stores one *order* of creature.

Within each pull-out drawer, there are file folders. Each folder is filled with one *family* of organisms. Inside each folder, you find a set of documents made of pages stapled together, like short magazines or school reports. Each report tells of a *genus,* a set of very similar living things. Each page of this report describes a different *species.* On our imaginary campus, humans have one page.

When you practice classifying animals, you will find it helpful to work from both the top down and the bottom up. You can do a lot of classification with common sense and observation.

In the twentieth century, scientists gained tools unavailable to earlier naturalists—the ability to look within the cells of organisms. Using the information stored in **DNA** and **RNA,** some classification systems based on domains were suggested. Domains are broader, higher-level categories than kingdoms. **Bacteria, Archaea,** and **Eukarya** are the names of divisions suggested by Carl Woese (b. 1928). Another system has only two kingdoms, animals in one and everything else in the other. Although we have come a long way since Linnaeus, his system is still in use.

# Vocabulary Words

| | |
|---|---|
| Archaea | group of single-celled organisms classified as prokaryotes (cells without nuclei) along with bacteria. Many forms live under extreme conditions. |
| Bacteria | group of organisms with genetic material that is not contained in a nucleus |
| binomial | having two names. In taxonomy, a two-part name consists of genus and species. |
| DNA | deoxyribonucleic acid; a double twisted helix inside a cell that contains the genetic code for that organism |
| Eukarya | set of organisms whose cells have a definite nucleus |
| hierarchy | the order of classification determined by how broad or narrow the groupings are |
| RNA | ribonucleic acid, a chemical active in transferring genetic information and building proteins |

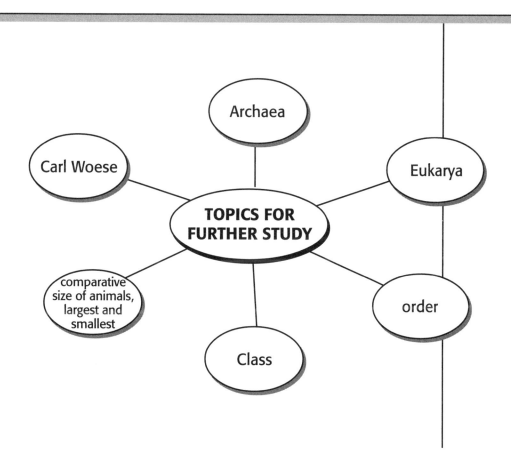

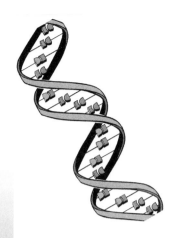

# CHAPTER 3
# Cells and DNA

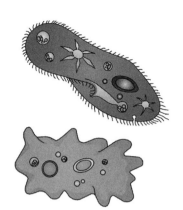

**TIME LINE**

| Year | Notable Event |
|---|---|
| 300s B.C. | Aristotle described the different function of the yolk and the white part of eggs. |
| 1665 | Robert Hooke observed cells in plants. |
| 1673 | Anton van Leeuwenhoek used his microscope to observe single-celled organisms and other previously invisible aspects of life. |
| 1831 | Robert Brown named the nucleus. |
| 1838 | Matthew Schleiden claimed that all plant materials are composed of cells. |
| 1839 | Theodor Schwann identified eggs as cells and theorized that all animal tissue is composed of cells. |
| 1847 | Ignaz Semmelweiss connected the practice of hand washing and slowing the spread of disease. |
| 1856 | Louis Pasteur wrote that yeast causes fermentation. |
| 1860 | Gregor Mendel experimented with the laws of heredity. |
| 1869 | Friedrich Miescher discovered DNA. |
| 1882 | Walther Flemming discovered chromosomes and cell division. |
| 1928 | Alexander Fleming used penicillin mold to kill bacteria. |
| 1944 | Oswald Avery linked DNA to heredity. |
| 1953 | James Watson and Francis Crick developed the double helix model of a DNA molecule. |
| 1967 | Thomas Brock discovered microbes living in hot springs in Yellowstone National Park. |
| 1983 | Luc Montagnier and Robert Gallo identified the HIV virus. |

**Materials per Team**

- diagrams or drawings of cells
- Jello® and candies, or colored clay, marbles, stones, or other small objects

# A Model Cell

## *Modeling a Cell*

Not all cells look alike. There are many types in the human body alone. Groups of cells working together form tissues, and groups of tissues make organs. So although cells are the universal building blocks of life, different functions are performed by different types of cells. However, there are enough similarities so that students can make a model containing the major parts of a cell.

## Activity

Essentially, a cell has three major parts: the membrane, cytoplasm, and organelles. First, decide whether you want to make model cells with edible materials. Edible cells will have candy pieces (representing organelles) embedded in light colored Jello® (representing cytoplasm), which you and/or the class will have to prepare a day or two ahead. Making this kind of model is somewhat like building a gingerbread house—the materials are edible but, after being handled, will not be appropriate to eat. So be sure to keep extra pieces of cellular matter on hand for snacking if the models are handled too much to eat.

To make inedible models, replace the Jello® in the following instructions with modeling clay or dough for the cytoplasm and colored clay, marbles, and other objects for the organelles. In fact, students can participate in finding and choosing objects to be organelles based on class research into cell structures.

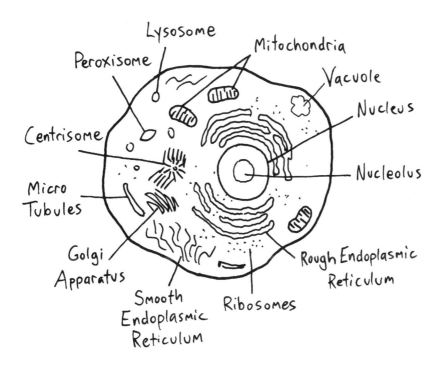

Each team will need the following items for its model:

▶ **nucleus:** Largest organelle, contains genetic material (DNA on chromosomes). Represent with a large marble, gumball, stone, or hard candy.

▶ **endoplasmic reticulum (ER):** Canals or tunnels between the cell membrane and the nuclear membrane. Represent with hollow licorice sticks, rolled-up "fruit leather," sticks, straws, or clay tubes. Make two to four per cell.

▶ **mitochondria:** Energy producer of the cell. Represent with bright-colored objects to show power production; small beads, small candies, clay bits, or tiny stones. Make two to four per cell.

▶ **Golgi apparatus:** Series of flat organelles that transport cell products. Represent with ribbon candy, fruit leather, or sticks of chewing gum. Make two per cell.

- ▶ **ribosomes:** Attach to and work on the ER to synthesize proteins. Some are also scattered throughout the cytoplasm. Represent with sprinkles, large salt crystals, glitter pieces. Use about $1/2$ teaspoon.
- ▶ **lysosomes:** Spherical bodies that contain enzymes that destroy foreign bodies. Represent with small "dot" candies, clay spheres, or stones. Make two per cell.
- ▶ **vacuoles:** Fluid-filled sacs that are the storage chambers of the cell. Represent with beans, jelly beans, raisins, or seeds. Make two to four per cell.
- ▶ **peroxisomes:** Spherical bodies that protect cells from toxins. Use small white spheres, four to six per cell.
- ▶ **centrosome:** Pair of microtubules that replicate and divide during cell division. Use thin licorice or other candy or string. Make one pair per cell.

Cut the Jello® into large, roughly cube-shaped pieces, giving one to each group. Or, you can make the Jello® in plastic cups, giving one to a group. Each group can cut their Jello® into two sections. They can then hollow out the center of one half and add the parts according to the diagrams. Be sure each team adds a key to their model so that viewers can identify the pieces and discover the function of the parts of a cell.

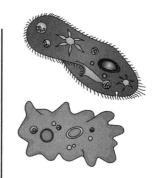

# READING:
# Cells: Units of Life

"Omnis cellula e cellula." This Latin phrase, formulated by Rudolf Virchow (1821–1902) in 1855, means "all cells arise from cells." In the mid-nineteenth century, German scientists (including Virchow) had determined that cells were the basic building blocks of life. Their theory contradicted the theory of **spontaneous generation** and led to many breakthroughs in understanding how all living things operate.

People knew about cells since at least 1665 when Robert Hooke (1635–1703) viewed **cork** through a microscope. Hooke compared the tiny compartment-like structures he saw in wood to the cells of monks, tiny cubicles in **monasteries** occupied by the religious workers. But understanding the importance of cells did not come for about another two hundred years.

Spontaneous generation was widely believed to be the way many forms of life began. If the conditions were right, the theory supposed, organisms could simply appear where there previously had been none. William Harvey (1578–1657), most famous for his work on blood circulation, disagreed and theorized that tiny eggs and seeds would be the cause of tiny animal and plant life. In 1668, Francisco Redi (1626–1697) used a controlled experiment to prove that rotting meat does not spontaneously create **maggots**—flies must lay eggs that later hatch. However, spontaneous generation proved to be a very persistent theory.

As microscopes improved, scientists gained increasingly clear views of tiny objects. The secrets of nature are often revealed in her details. In 1831, Robert Brown (1773–1858) noticed small objects inside cells. Brown named this body the *nucleus.*

In 1838, Matthias Schleiden (1804–1881) stated that all plants are made of cells. The next year, Theodor Schwann (1810–1882) proposed the same theory for animals. Schwann claimed that eggs are actually cells and that life begins from a single cell.

In the next decade, Virchow theorized that a disease is caused when an organism's cells get out of control. When Louis Pasteur (1822–1895) later championed the germ theory, Virchow's theory needed to be refined. Pasteur attributed disease to invasion by foreign bodies. It turns out that both theories may be true. Out-of -

control cells are a symptom of **cancer** and other illnesses. Many diseases are caused by the invasion of **bacteria, viruses,** and other tiny organisms.

The cell is a remarkable piece of nature's engineering. The basic structure of an animal cell is similar in creatures large and small and has been in use for millions of years. The way things work on a **molecular** level in complex animals is similar to the behavior in single-celled creatures and other simple organisms. Because of this universal design, scientists can experiment and study cells in a variety of animals to learn about the human body.

The component parts of a cell were developed by living organisms long ago but remained hidden from human view until the microscope was invented. If you build a model cell, think about how complicated and tiny a real cell must be. You will appreciate how difficult it must have been for the pioneer biologists to unlock its secrets. In the future, new tools will help scientists learn even more about cells, microscopic keys to life.

## Vocabulary Words

bacteria ......................................... group of organisms with genetic material that is not contained in a nucleus

cancer ........................................... general name for invasive malignant growths of cells

cork ............................................... plant tissue in outer layer of stem that prevents drying out. The cork of a certain oak species is used for bottle stoppers and other applications.

maggot .......................................... general name for the larval stage of some insect species, especially flies

molecular ....................................... referring to individual molecules

monastery ...................................... home for religious people

spontaneous generation
theory ........................................... idea that organisms may arise from non-living things, sometimes called *abiogenesis*

virus .............................................. tiny parasite composed of DNA or RNA and protein

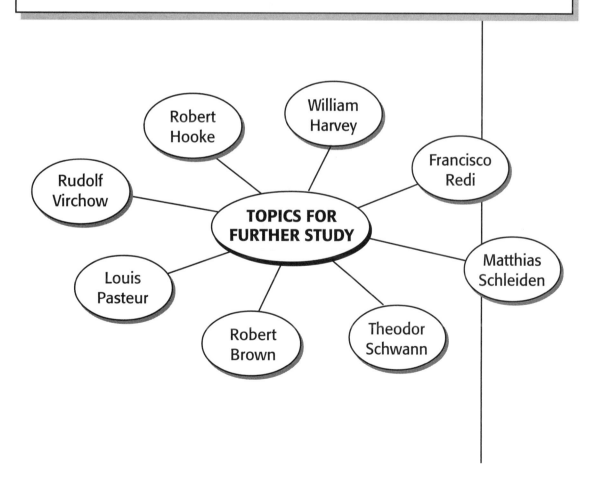

From *Science Giants: Life Science* © Good Year Books. This page may be reproduced for classroom use only by the actual purchaser of the book. www.goodyearbooks.com

# Genetic Inheritance

## Mendel's Peas

In this activity, students will simulate Gregor Mendel's famous experiments. Mendel discovered that the mechanism of inheritance depends on the contributions from two parents and follows a mathematical relationship. He looked at a variety of characteristics, but in this activity students will consider the shape of seeds, either spherical (S) or dented (d).

### Activity

Begin by explaining to students that genes come to offspring in pairs, one from each parent. Each gene is chosen by chance from the parent's pair of genes. If a plant self-pollinates, the resulting offspring only has one parent. But the offspring still gets two genes, one each from two identical pairs.

The spheres (marbles or clay balls) and aluminum foil balls represent the two kinds of genes in this activity. Pairs of genes will be placed in plastic bags. There can be three kinds of pairs: two spheres, two foil balls, or one sphere and one foil ball.

Have students place two spheres in a bag. This represents a pea plant that carries genes only for producing spherical-seeded offspring. Draw it on the chalkboard and label it "S/S." Put two dented foil balls in a bag and label it "d/d." This represents a pea plant carrying genes programmed to produce only dented seeds. As Mendel did, students will cross-pollinate these plants.

Show how to set up a grid called a Punnett Square (see page 57) to check the outcome of crossing the plants. Place parent plant 1's gene pair across the top and parent plant 2's gene pair along the left side. Fill in the squares as you would an addition chart.

All four possible combinations pair a gene for spherical seeds with a gene for dented seed. Mendel found that all the offspring grew spherical seeds. *S* genes are dominant over *d* genes, meaning that a plant inheriting the *S/d* combination will produce spherical

## Probability Chart for Generation 2

**Parent 1's Genes**

| Parent 2's Genes | | S | S |
|---|---|---|---|
| | d | S/d | S/d |
| | d | S/d | S/d |

seeds. So each of these four offspring would produce spherical seeds but would carry genes for dented seeds. The probability for spherical seeds in the generation described by the chart is 100%. Tell students they are going to simulate what might happen to future generations, recreating a famous experiment Mendel performed.

He grew a plant from the generation of offspring carrying the *S/d* gene pair and had it self-pollinate. Set up another grid to learn what could happen in that circumstance. (See "Probability Chart for Generation 3" on page 58.) The top and side labels will both be *S* and *d*, because one plant is contributing two sets of genes. This time, have students flip a coin to choose a gene from each pair. Heads can indicate *S*, tails *d*. Record the outcomes as each team of students finds four offspring.

| | Gene 1 | Gene 2 | Result |
|---|---|---|---|
| Offspring 1 | S | d | Smooth |
| Offspring 2 | d | d | Dented |
| Offspring 3 | | | |
| Offspring 4 | | | |
| | | | |
| | | | |
| | | | |

57

When students do simulation plant crossings and flip a coin to choose one or the other of the genes in each pair, don't remove the genes from the bags. Reproductive cells produce copies of the organism's genes to pass to offspring.

Mendel found that the ratio of *S* to *d* in the second generation was 3:1. Because probability is involved, outcomes will vary, but class results should come close to the theoretical probabilities with a large enough sample. Mendel had a very large sample in his experiments, and some historians of science think he may have manipulated results to suit his theory! But scientists now understand much more about the mechanisms of heredity and realize why the math works. The table for Generation 3 explains the 3:1 ratio.

### Probability Chart for Generation 3

**Parent 1's Genes**

| Parent 2's Genes | | *S* | *d* |
|---|---|---|---|
| | *S* | *S/S* | *S/d* |
| | *d* | *S/d* | *d/d* |

Encourage the students to try different crosses of the peas and see how the results work out. Coin flipping isn't really necessary when the bag contains two identical genes in the pair.

Using the grid method to predict results and coin flipping to create actual data, students will understand the mathematical basis of genetics as Mendel discovered. This simulation isolates one characteristic and is therefore an oversimplification, but students will appreciate the basic idea of genetic inheritance.

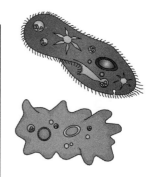

## READING:
# From Parent to Offspring

Charles Darwin (1809–1882) was renowned for developing one of the most important ideas in life science in the nineteenth century. Darwin explained the theory of **evolution,** which states that individual organisms are selected by nature in a life-or-death test—if they succeed, they reproduce and more organisms like them are born. If they fail, they don't reproduce and do not pass along the characteristics of their bodies. A whole population may dwindle when too many individuals fail to reproduce.

Darwin believed changes and characteristics are passed along from one generation to the next. But neither he nor anyone else knew how. A monk experimenting with peas in his garden discovered some of the first important clues about **genetics,** but even he didn't realize how important his work would be.

Gregor Mendel (1822–1884) was an Austrian monk who was the son of a farmer. During his experiments with peas, Mendel recorded the frequency with which plant **offspring** inherited characteristics from their parent plants. In 1865, he had a paper published by a local science group, but the news of his work did not spread.

What did Mendel discover? His experiments revealed important concepts about how traits can be passed along from generation to generation. Plants can receive **genes,** carriers of information about how to grow, from either one or two parents.

Some characteristics they inherit are of an "either/or" variety, operating like a switch. There is no middle ground—something about the plant is one way or another. Mendel's peas were either smooth or dented.

Other characteristics plants inherit cover a range of attributes. For example, flower colors could cover a spectrum from white, through shades of pink, and on to red. Think of a light switch again. Some bulbs simply turn on and off, while others can vary in brightness because of a dimmer switch, or **rheostat.**

When Mendel experimented with the "either–or" characteristics of peas, he found that there were definite patterns to **inheritance.** When he crossed, or used as parent plants, a tall plant and a short plant, all the offspring were tall. In the next generation, about three-fourths were tall and one-fourth short.

Mendel reasoned that either tallness or shortness would be inherited. Each parent contributes one factor, which were later named *genes*. Of the offspring's two inherited genes, how does one get chosen and go on to determine the height of the new plant?

Mendel figured out that one trait (tallness) would be dominant, or always chosen over the other. The recessive trait (shortness) would be inherited but would not show up in the new plant. When two shortness genes were inherited, the offspring would be short. But as long as at least one of the two inherited genes carried a program for tallness, the new plant would be tall. One of the most famous characteristics Mendel studied was the shape of pea seeds.

After he had performed years of experiments and recorded his results, Mendel wrote to scientists to share his work. He received no encouragement or praise and eventually gave up his genetic investigations.

By 1900, other scientists reached the same conclusions as Mendel and learned about his experiments. Mendel's achievements were noted and history celebrates him as a pioneer, the discoverer of the basic laws of genetics.

Darwin never found out the mechanics of heredity that Mendel and those who followed discovered. Since the discovery of **DNA** and other advances in the second half of the twentieth century, much of the work of life scientists takes place on a **molecular** level, where the genetic information is stored. But the study of genetics got started in the vegetable garden of an eighteenth-century monk.

## Vocabulary Words

DNA .......................................... deoxyribonucleic acid; a double twisted helix inside a cell that contains the genetic code for that organism

evolution ................................... the theory that genetic changes from generation to generation over time cause species to change gradually

genes ........................................ sections of DNA that produce a trait in an organism

genetics .................................... the study of heredity

inheritance ............................... characteristics transmitted by genetic material

molecular ................................. referring to individual molecules

offspring ................................... descendants of organisms—for example, the children of human parents

rheostat ................................... an electrical resistor that can regulate current in a circuit. Dimmer switches are a type of rheostat.

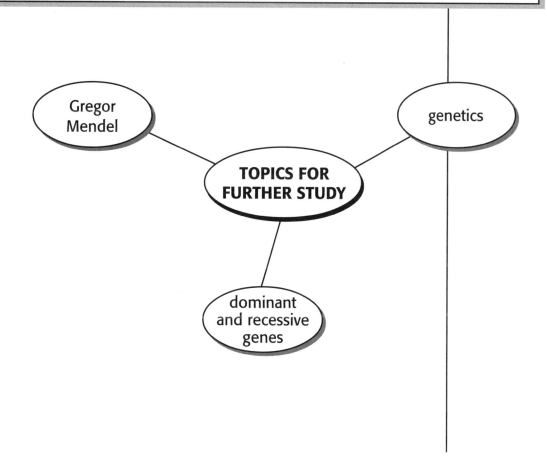

From *Science Giants: Life Science* © Good Year Books. This page may be reproduced for classroom use only by the actual purchaser of the book. www.goodyearbooks.com

## Materials per Team

- index cards or sturdy paper cut into small rectangles
- several packages of two colors (to represent sugar and phosphate groups) of flexible licorice candy sticks, cut into shorter lengths
- toothpicks
- Styrofoam or other packing "peanuts," or soft candy pieces like gumdrops or marshmallows (four colors to represent nitrogen bases)

# Model DNA

## *The Ladder of Life*

This activity models the structure of DNA, the type of protein molecule that carries the code telling living cells how to behave. Students will first simulate the pairing mechanism that allows DNA to replicate. Then they will construct a "double helix," the now famous shape discovered in the 1950s. The building process might be a little tricky and difficult to manage, but students will learn how DNA replicates itself.

## The Rungs of the Ladder

To prepare for this activity, label sturdy paper rectangles or index cards (may be cut in half or even quarters for economy) with the names of the chemical bases that form the rungs of the DNA ladder. These bases always combine in pairs.

**guanine (G) <—> cytosine (C)   adenine (A) <—> thymine (T)**

Cut each type of base so that it connects to its partner like a pair of puzzle pieces. Be sure the connecting parts are different enough so that matches can only be made between G and C and between A and T.

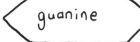 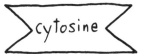  

Have students lay six to ten of the cards on a table in a vertical column to represent a sequence of bases. Using other cards, have them match each base in the sequence to create a double column of connected bases. Be sure they match and connect G and C bases and A and T bases.

These nitrogen-containing bases are just one part of the DNA molecule. Students will represent the other parts of DNA in the model they'll soon make by building entire nucleotides, but the bases are the parts that connect. DNA molecules are made up of many nucleotides. Each is a three-part unit, containing a nitrogen base, a sugar, and a phosphate group. When the nucleotides are assembled, they assume a form like a twisted ladder.

Nitrogen bases from two different nucleotides link together to form the rungs. Each base is attached to a sugar that is in turn attached to a phosphate group. The sugars and phosphate groups on the ends of the bases form the side rails of the ladder.

Next, have students carefully separate the two columns of bases, and add a match onto each newly separated partner. Students should see that two copies have been produced. Of course this is very simplified, but it demonstrates the essential idea of replication.

## Putting the Ladder Together

To make the model, give every group of students three pieces of each color of licorice candy to build one side rail. Choose one color to represent sugar and one to represent phosphate groups. Students should alternate the colors, connecting the candy with toothpicks.

Next, stick another toothpick into a piece of soft candy. Stick the other end of the toothpick into a sugar on the side rail. The soft candy represents a nitrogen base. Repeat until all three sugars have attached nitrogen bases sticking out.

Each team now has a "triplet." This set of three bonds can combine with other triplets to form an amino acid, a part of a protein. Here's where the coding comes in. Remember, the four base groups can only combine in specific pairs.

**guanine (G)<—> cytosine (C)  adenine (A) <—> thymine (T)**

Be sure to designate different colors of candy to represent each of the four bases.

Have students compare triplets with other groups. Do any match at all three bases according to the G<—>C and A<—>T rule? Most likely new triples will have to be built to provide complementary matches for the first set. That will model how DNA replicates. The strands of the double helix separate and eventually connect to a new partner, producing two DNA molecules.

This may seem painstakingly detailed to students. Within a cell, any mistakes in sequencing the bases or the amino acids (from the triplets) will cause a defect in the long protein string that forms the DNA and could damage or kill the organism.

Much of the work of modern biochemistry and related fields takes place at the molecular level with DNA. As students learn more about the cell's genetic machinery, they should be better able to follow the unfolding discoveries in life science and appreciate how far our knowledge has come.

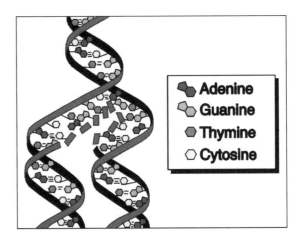

### READING:
# Discovering the Structure of DNA

There are many legends in the history of science. Entertaining stories tell how Archimedes, Galileo, Newton, and other giants made discoveries that changed the way we see the world. Because it has been many generations since the events took place, there is often more legend than fact to the tales.

In contrast, the story about how **DNA's** structure was discovered is well documented. In the middle of the twentieth century, groups of scientists were racing to be the first to figure out the mystery of life's code. Competition was intense. Discovering how DNA operated would be a remarkable breakthrough. Curing diseases in new ways, developing stronger characteristics in useful species, avoiding weaknesses that threaten cells—the prospects for change were awesome. Surely the team or individual who solved the puzzle of DNA's structure would become famous.

DNA was discovered in 1869 by Friedrich Miescher (1844–1895), a Swiss scientist. Miescher extracted DNA from various animal cells. He identified it as an **acid,** but the real significance of DNA would be discovered after his death. Like Gregor Mendel (1822–1884), Miescher never knew how crucial his contribution toward understanding **heredity** and cell function would be.

In the 1940s and 50s, scientists became sure that DNA controlled heredity. Thomas H. Morgan's (1866–1945) work had pinpointed genes as the elements of heredity. Morgan worked with fruit flies and learned that some sets of inherited characteristics are linked together.

Mendel's laws of heredity worked because of **genes.** Oswald Avery (1877–1955) and his co-workers proved by isolating DNA that it was a major component of genes and the true controller of heredity.

To understand how DNA worked, scientists needed to solve the riddle of its chemical structure. Linus Pauling (1901–1994), an American chemist, proposed the idea of a **helix,** a three-dimensional spiral. Special photographs called **X-ray diffractions** taken by Rosalind Franklin (1920–1958) and Maurice Wilkins (1913–2004) in England supported the helix theory.

But every attempt at building a model failed due to some flaw. In 1953, Francis Crick (1916–2004) and James Watson (b. 1928) built a model of DNA as a *double* helix, which turned out to be correct. Crick and Watson are famous for their discovery, which they built upon the work of others—another example of seeing farther from the shoulders of giants!

Making a model of DNA took lots of imagination and scientific detective work. The elegance of its design is truly a key to the life of the cells and, therefore, all living things. To discover the structure of DNA, scientists needed knowledge of biology, chemistry, and physics. While scientific training has become more and more specialized, advances in several fields often combine to spark the discoveries that cause the greatest impact. The amount you can learn in one class period building a model is the result of hundreds of years of work.

DNA molecules are made up of many **nucleotides.** Each is a three-part unit, containing a **nitrogen base,** a **sugar,** and a **phosphate** group. When the nucleotides are assembled together in a DNA molecule, they assume a form like a twisted ladder.

Nitrogen bases from two different nucleotides link together to form the rungs or steps across the ladder. There are four types of bases. Each type can only connect to one specific other type.

The other side of each base is attached to a sugar that is in turn attached to a phosphate group. The sugars and phosphate groups on the ends of the bases form the side rails of the ladder.

Sets of three "rungs" called *triplets,* or **codons,** can combine to form **amino acids,** a part of a **protein.** The proteins essentially do the work of the cell.

Understanding the way DNA works has caused a revolution in life science and the amount of information is multiplying rapidly. As you continue learning science, you will know more about cells than anyone knew during Darwin's life.

# Vocabulary Words

acid .......................................... sour-tasting liquid containing hydrogen ions (H⁺)

amino acid ............................ molecules crucial to the formation of DNA

base .......................................... substance that neutralizes acids, removing the hydrogen ions to form water

codon ....................................... a triplet of nucleotides in a DNA molecule that spells out or specifies a particular amino acid

diffraction ............................... bending of waves around an obstacle

DNA ......................................... deoxyribonucleic acid; a double twisted helix inside a cell that contains the genetic code for that organism

genes ........................................ sections of DNA that produce a trait in an organism

helix ......................................... three-dimensional spiral

heredity ................................... process by which traits are passed from one generation to another through genetic information

nitrogen .................................. gas that makes up almost 80% of the atmosphere

nucleotide ............................... segment of DNA consisting of a base, a sugar, and a phosphate

phosphate ............................... certain chemical compounds containing phosphorus and oxygen

protein .................................... specific chain of amino acids that is vital to enzymes and cells

sugar ........................................ general term for some large carbohydrates

X rays ...................................... radiation of a specific high-frequency wavelength in which the photons have high penetrating power and can pass through solid objects

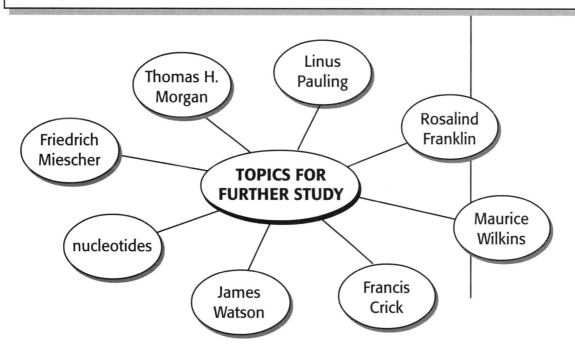

From *Science Giants: Life Science* © Good Year Books. This page may be reproduced for classroom use only by the actual purchaser of the book. www.goodyearbooks.com

## Materials per Team

- large book (such as an encyclopedia)
- worksheet on page 70
- calculators (optional)

# The Code of Life

## *I Can Read You Like a Book*

The information to create a new organism is coded within each cell's DNA. How can something as complex as a sunflower or a human being be programmed in microscopic form? This simulation will help students visualize the structure and hierarchical mathematics involved in cell reproduction and genetics.

## Activity

Display or pass out samples of textbooks or reference books. As students look at a page, have them examine the text and compute how many letters, words, and lines there are per page. Distribute the handout on page 70 and instruct students to calculate the number of letters in the book, filling in their own numbers on each line. If you decide to agree upon some round figures to use collectively, here are some sample numbers:

|   |   |
|---:|:---|
| 10 | words in a line of text |
| x 2 | columns per page |
| 20 | words across the page in a line |
| x 50 | lines down the page |
| 1,000 | words on a page |
| x 6 | letters per word |
| 6,000 | letters per page |
| x 500 | pages in a book |
| 3,000,000 | letters in a book |

After students have completed their estimates of these figures and discussed their findings, they can begin to build an analogy of how each molecule of DNA contains a virtual library full of information about how an organism is built. DNA molecules are long strands of linked chemical modules called *nucleotides*. The nucleotides contain sugars, phosphates, and a base molecule. There are four kinds of base modules and they connect in pairs; adenine (A) and thymine (T) bond together, and cytosine (C) and guanine (G) do also. In our analogy, base pairs A<—>T and C<—>G are the "letters" and the order in which they line up forms the "words."

Here come some big numbers. One set of human chromosomes contains DNA molecules believed to made up of about three billion base pairs. By our analogy, the base pairs are equivalent to three billion "letters." Divide by six letters per word and we need five hundred million (500,000,000) words. At this point, be sure students understand that molecular strands ("words") can contain many bases, but our analogy divides them into the average size of words.

At 1,000 words per page, our library needs five hundred thousand (500,000) pages to hold the five hundred million (500,000,000) words. If each book has 500 pages, we need 1,000 books to contain the information in a set of human chromosomes.

After all that math, students should appreciate how much information is contained in each cell of their bodies.

# I Can Read You Like a Book

Words per line of text            _____

Columns per page            x  _____

Words across the page per line            _____

Lines down the page            x  _____

Words per page            _____

Letters per word            x  _____

Letters per page            _____

Pages per book            x  _____

Letters per book            _____

## Number of base pairs in 3,000,000,000 human chromosomes ("letters")

Letters per word            ÷  _____

Number of words            _____

Words per page            ÷  _____

Number of pages            _____

Pages per book            ÷  _____

Number of books            _____

# Genetic Engineering

## *Patched Genes*

After students simulate how DNA strands carry information coded as sequences of molecules, in this activity they can demonstrate how scientists are learning to manipulate genes.

### Activity

Give each team of students five lengths of different-colored yarn as described above. We'll call the longer lengths A, B, and C and the shorter ones D and E. Have the teams cut A, B, and C into a few approximately equal lengths and cut D and E in half.

Next, take one length of D and tie both ends to the ends of a section of A, creating a loop. Place the loop in one of the larger containers and label the container "#1." Tie a length of E to another section of A and place that loop in the container's twin, labeling it "#2." Put the unused sections of D and E into the smaller jar.

## Materials per Team

- five lengths of different-colored pieces of yarn— three about 50– 100 centimeters (20–40 inches) and two about 25 centimeters (10 inches)
- two identical large containers and one smaller jar
- tape, if students have difficulties tying knots
- scissors

Tie some of the sections from the remaining cut pieces of A, B, and C into loops. Match them into pairs and place one of each pair into each of the twin containers. These loops can vary in length and number of colors but must be in pairs. Leftover sections of yarn without a match can be placed in the smaller jar. Students are now ready to simulate genetic engineering, or recombinant DNA technology.

In this model, yarn represents genes and other molecular sequences along strands of DNA. Each complete loop stands for a chromosome, the structures in the cell that carry DNA. The large containers represent cells from the same or very similar species, and the smaller jar represents a bacterium into which genes have been inserted.

Label one of the large containers "normal," designating it as a cell with a normal contingent of genes. Label the other large container "defective," meaning it contains a defective gene. Students will have to find the defective part of the chromosome in container two and repair it.

Switch sets of bottles among teams so that each group receives the twin containers that another group prepared. Students will have to find the defective, or undesirable, gene by comparing the chromosomes in cell #1 to the chromosomes in cell #2 and finding the loops that are different. By using scissors to cut the defective section and tying yarn from the bacterium (smaller jar), can they remove the undesirable gene in cell #2 and replace it with the target gene so it matches cell #1?

Scissors do the job of breaking the strands that enzymes (molecules that stimulate a reaction among other molecules) or viruses do in the laboratory. The bacterium holds the section of DNA that will function correctly in the cell.

Another scenario for genetic engineering involves "improving" a species. Scientists may want to insert a desirable gene from one species into the chromosomes of another species. For example, a gene from another plant that helps it tolerate drought might be inserted into corn to make the corn more resistant. This procedure is controversial. Changed species are referred to as being "genetically modified" or GM species.

Students may find making the repairs to the genes difficult. Real genetic research takes years! Students may have to borrow "genetic material" from another team to complete their work. Encourage them to find out what kinds of genetic engineering are already possible and what seems to be on the near horizon.

## READING:
## THE CODE OF LIFE *AND* GENETIC ENGINEERING
# *How the Code Is Written*

All living things are made of cells. Some organisms, in fact, are *only* one cell. Most cells are too small to see with our unaided eyes. Since the invention of the microscope, scientists have been studying the cell to learn more about how it works.

Robert Hooke (1635–1703) observed and recorded the structure of cork in 1665. In the nineteenth century, scientists came to the conclusion that all living things consist of cells. In the 1830s, Robert Brown (1773–1858) named a part of the cell the **nucleus.**

Cells with a nucleus are called **eukaryotic,** and those without a true nucleus are called **prokaryotic.** The prokaryotic cells tend to be smaller and appear less complicated. All cells have some kind of outer covering, either a wall or a membrane. Within the cell the processes necessary for life are occurring continuously. Fuel is received, energy produced and used, chemicals processed, products and waste produced, messages received and sent, and new cells made. Those are only some of the jobs that a cell does.

**Chromosomes** can be considered the information processors of the cell. Chromosomes form from **DNA molecules** bonding to **proteins.** DNA molecules are long strands of chemical modules, or groups, linked together.

The section of DNA that establishes a code for reproducing itself is called a **gene.** The order of the bases along the DNA strand creates a code that determines which proteins will be made.

Barbara McClintock (1902–1992) learned that genes could move position on chromosomes. She called this action "jumping genes" and her many discoveries using corn plants helped set the stage for the revolution in genetics in the late 1900s.

The chemical structure of the cell's chromosomes directs the cell's activities and determines its future. The long strands of DNA, which are coiled around protein molecules, have been discovered to have a double **helix** structure resembling a twisted ladder.

The rungs of the ladder are combinations of nitrogen-containing bases. The bases can only bind to one other kind of base. They combine in units of three to make a specific code for the production of amino acids. The rungs "unzip" down the middle where the bases connect and then attach to other bases, creating exact replicas of themselves. Once this process of nature using a copying code was understood, more could be learned about genetic changes, or **mutations.** In the late years of the twentieth century, scientists began to unravel the code. They learned which parts of the DNA coded for some outcomes. That knowledge opened up the possibility of repairing and modifying the actions of the cell.

Repairing a defective part of the DNA code to cure a disease in someone would be a brilliant medical breakthrough. But some people worry that modifying the code to create changes in an organism for the benefit of humans may present problems.

If **genetically modified (GM) species** spread into the natural environment, they could overwhelm native species. The GM organisms might outcompete the local organisms and drive them to extinction through natural selection. Beneficial traits in the native species might be lost. This controversial subject will be debated more strenuously as scientists become better able to modify the genes of organisms.

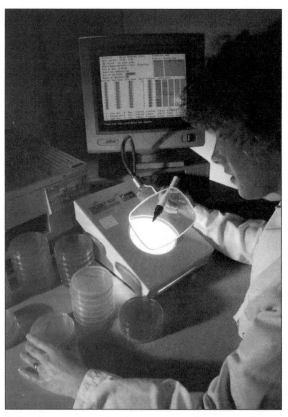

Scientists are learning how to experiment on the structure of DNA within cells.

## Vocabulary Words

chromosomes ........................ structures in cells that contain genetic information

DNA ........................................... deoxyribonucleic acid; a double twisted helix inside a cell that contains the genetic code for that organism

eukaryotic ............................... cells in which the genetic material is inside a nucleus

genes ........................................ sections of DNA that produce a trait in an organism

genetically modified (GM)
species ..................................... organisms whose genetic material has been deliberately altered

helix .......................................... three-dimensional spiral

molecule .................................. two or more atoms bonded together

mutation .................................. permanent change in the genes of an organism

nucleus ..................................... in a cell, the part separated by a membrane and containing most of the cell's DNA. In an atom, a positively charged mass containing protons and neutrons.

prokaryotic .............................. cells in which the genetic material is not inside a nucleus

protein ...................................... specific chain of amino acids that is vital to enzymes and cells

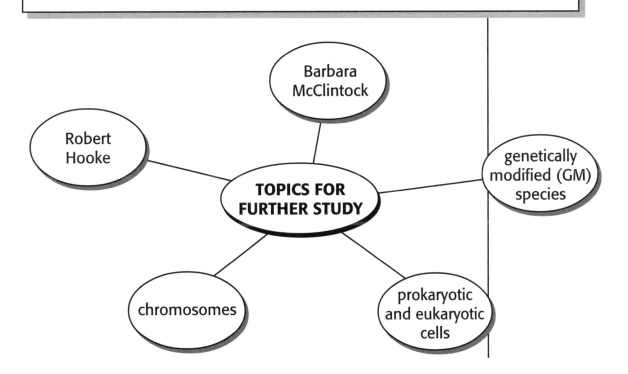

## Materials per Team

- locking plastic bags
- dry powdered baking yeast
- sugar and other sweeteners
- water
- thermometers to measure water temperature
- magnifying lenses
- variety of foods

# Yeast and Molds

## *Growing Yeast and Molds*

Growing yeast and molds helps build an appreciation for the variety of small life-forms all around us. As students design their own experiments, encourage them to discover the circumstances that allow these organisms to flourish. Emphasize the importance of controlling variables.

### Activity

In this activity, each team will devise an experiment in which yeast is grown under different conditions. However, first let them establish a baseline set of parameters as a control by placing one teaspoon of yeast, one teaspoon of sugar, and a cup of warm water into a large locking plastic bag. They should record the initial water temperature.

This activity should create a substantial colony of yeast fungi. After a specified amount of time, the circumference of the bags can be measured to record some growth data. As yeast eat and grow, they produce carbon dioxide ($CO_2$), which puffs up the bags.

Which condition would the students want to change to determine its effect upon the yeast's growth? Suggestions should include initial water temperature, temperature around the bag during the experiment, water without sugar, more or less sugar, other sweeteners in place of sugar, and so on. Does it matter if the yeast is grown in the darkness or light? Remember to keep variables constant except for the one being tested, including water temperature. Try out the experiments and measure the bags to find out how puffed up they became.

Growing molds can proceed in much the same way. Dampen some foods and zip them into plastic bags. Vary the conditions using our approved scientific techniques. Wet versus dry, sunny location versus dark, refrigerator versus warm area, and so on. Look at the molds with magnifying lenses while still in the bags. Density of growth can be used for collecting data, as the molds will not puff up the bags with $CO_2$. Have students estimate and record the percentage of surface area covered by mold. Caution students not to touch the molds, as they may be harmful and should be left in the bags. To be safe, dispose of all molds and yeast growths after the activity.

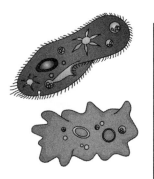

# *READING:*
# *Tiny Life-forms*

Humans, and most animals, use **oxygen** to obtain energy from food. This process, called **aerobic respiration,** depends on our breathing.

We take oxygen from the air and combine it with **glucose** from our food. Glucose is the main sugar produced by living cells and the form our food takes for delivery to our cells. The resulting reaction creates **carbon dioxide** (which we breathe out), water (which makes up much of our body mass), and energy.

**Yeast** organisms can release energy from food a different way. They do it without using oxygen. In that process, called **anaerobic respiration,** yeast cells convert glucose to carbon dioxide, **alcohol,** and energy. When bread dough rises, carbon dioxide bubbles make it puff up. Baking kills the yeast cells and the bread stops rising.

For centuries, people have used yeast for bread baking and in the brewing of wine and beer. Alcohol brewing is called **fermentation.** But even though yeast has been used for centuries, it was not until the 1830s that some scientists, including Theodor Schwann (1810–1882), became convinced that yeast are living organisms. Along with Matthias Schleiden (1804–1881), Schwann popularized the theory that all living things are composed of cells. The famous French scientist Louis Pasteur (1822–1895) spent some years of his varied career working to help the wine industry of his country. He observed that yeast cells caused fermentation while they were living.

Modern discoveries about yeast continue to be useful by providing people with an expanding range of services. In the 1980s, vaccines produced by yeast were made available to help protect people from **hepatitis,** a disease of the **liver.**

Another member of the **fungus** kingdom can destroy our bread-baking accomplishments—**mold.** Mold spores travel through the air, land on dead organic materials (like bread), and make their own food from their hosts' resources. But the same group of organisms that spoil bread have provided a great boon to humanity.

In 1928, Alexander Fleming (1881–1955) discovered mold growing in his lab and he didn't realize then how important that would be. The uninvited mold that grew on Fleming's lab cultures killed some **bacteria** he was studying. Fleming named the mold **penicillin** in his report. Years later, Howard Florey (1898–1968) and Ernst Chain (1906–1979) developed a method for producing large quantities of this valuable **antibiotic.**

A variety of antibiotics have been produced as research continued. Bacteria grow and evolve very quickly to resist antibiotics. New forms of antibiotics must be developed continually to fight bacteria in this biological **arms race.** Scientists warn that we may lose the protection of antibiotics if we overuse them and encourage the evolution of dangerous new strains of bacteria.

## Vocabulary Words

| | |
|---|---|
| aerobic | process requiring oxygen |
| alcohol | group of organic compounds containing carbon, hydrogen, and oxygen atoms (OH) |
| anaerobic | describes a process that does not require oxygen |
| antibiotic | substance that kills bacteria |
| arms race | an escalation or weapons competition. An advance by one side triggers a buildup reaction from the other, prompting even more weapon stockpiling by the first side, and so on. |
| bacteria | organisms with genetic material that is not contained in a nucleus |
| carbon dioxide ($CO_2$) | common gas composed of molecules with one carbon atom and two oxygen atoms |
| fermentation | process performed without oxygen, breaking down compounds into simpler forms, usually producing alcohol |
| fungus (pl., *fungi*) | organism from the kingdom Fungi. This kingdom includes mushrooms, yeast, molds, and other organisms. |
| glucose | common sugar composed of molecules of carbon, hydrogen, and oxygen atoms |
| hepatitis | disease of the liver |

## Vocabulary Words *(continued)*

| | |
|---|---|
| liver | the organ in the body that works to form blood and to metabolize food |
| mold | name for certain organisms in the Fungi phylum |
| oxygen | gas that makes up about 21% of Earth's atmosphere |
| penicillin | antibiotic derived from the penicillium mold |
| respiration | breathing, or in cells, the use of oxygen |
| yeast | certain unicellular fungi |

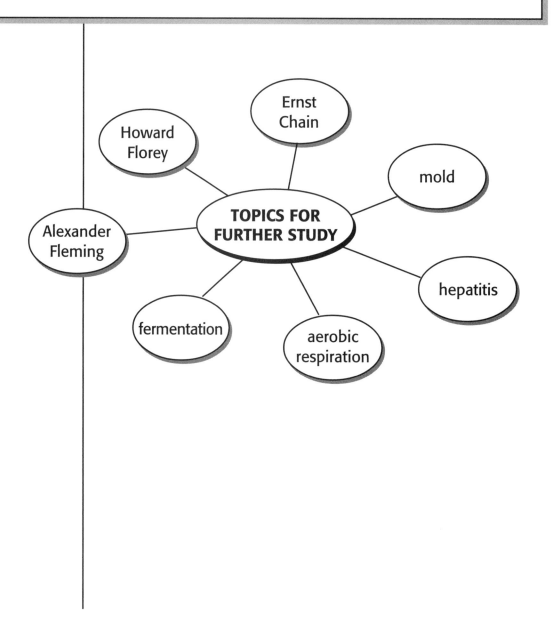

# CHAPTER 4
# Plants

From *Science Giants: Life Science* © Good Year Books. This page may be reproduced for classroom use only by the actual purchaser of the book. www.goodyearbooks.com

## TIME LINE

| Year | Notable Event |
| --- | --- |
| 300s B.C. | Aristotle observed and classified plants. |
| 1667 | John Ray classified plants based on the leaves within seeds. |
| 1675 | Marcello Malpighi published a plant anatomy book. He was an early user of microscopes. |
| 1682 | Nehemiah Grew published detailed descriptions of plant anatomy. |
| 1727 | Stephen Hales published a book explaining how plant structures operate. |
| 1737 | Carolus Linnaeus classified thousands of plant species. |
| 1772 | Joseph Priestley found that plants can make air breathable. |
| 1779 | Jan Ingenhousz discovered plants' need for sunlight and their use of gases. |
| 1804 | Nicholas de Saussure proved that plants require $CO_2$ from the air. |
| 1837 | Henri Dutrochet demonstrated the function of chlorophyll. |
| 1864 | George Washington Carver developed many uses for plants. |
| 1865 | Julius von Sachs published a botany textbook describing plant structure and function. |

## Materials per Team

- water plants (pond weeds)
- jars
- wide bowls
- water
- plastic bags
- house plants
- black paper
- scissors
- tape

# Photosynthesis

## *Observing Photosynthesis*

Plants make food using the chlorophyll in their leaves and the energy of sunlight. This process, called *photosynthesis*, can be written as an equation with the arrow sign meaning "yields":

**carbon dioxide + water ——> glucose (a sugar) + oxygen + water**

Written in chemical symbols and balanced mathematically, it looks like this:

$$6CO_2 + 12H_2O \longrightarrow C_6H_{12}O_6 + 6O_2 + 6H_2O$$

Notice how the number of atoms of C (carbon), O (oxygen), and H (hydrogen) on one side of the arrow equal the number on the other side.

## Activity 1

Students can observe some parts of this chemical activity. Have them place the pond weeds in a bowl of water. If you don't live near a pond or stream, buy a small quantity of water plants at a pet store. Students hold a small glass jar under water. Then they let the air escape from it by tipping it and letting the air bubble up to the surface. They should place the jar over the plants. Then they should place the bowl in a sunny location, and bubbles will rise in the jar. These bubbles are oxygen ($O_2$), a component of air.

---

## Activity 2

To demonstrate another product of photosynthesis, have students place a dry plastic bag around a potted house plant. They should wrap the open end of the bag around the stem and close with a twist tie. Then they can put the plant in a sunny spot and observe how the inside of the bag collects some of the water ($H_2O$) given off by the leaves.

---

## Activity 3

Students should cut out two small pieces of black paper. They then tape the paper around a section of a leaf of the house plant like a sandwich. They should check the covered section periodically to determine if the deprivation of sunlight is causing photosynthesis to stop occurring there.

Photosynthesis goes on all around us. Students have glimpsed a small part of the work plants do, work that is vital to our survival. Plants provide food, oxygen, and cooling as they evaporate water into the air.

### READING:
# Learning How Plants Produce Food

Plants can face very tough odds when it comes to survival. They need water, air, sunlight, structure and strength to hold themselves up, protection from predators, a way to reproduce . . . the problems seem endless. Too much sun and heat can shrivel them. Too much water can drown them. It isn't easy being green.

Scientists have studied plants for thousands of years and continue to be amazed at their versatility and durability. One of the first plant scientists we know about was Theophrastus (372–287 B.C.), a Greek who studied with Aristotle (384–322 B.C.). Theophrastus wrote *The Natural History of Plants* around 300 B.C. He described plant parts and behavior so clearly that his work was studied well into the Middle Ages. He was interested in plants from other regions as well as those of Greece.

When the **Renaissance** sparked European interest in science, **botany** began to grow. Carolus Linnaeus (1707–1778) systematically named thousands of plants by assigning a two-part name for each different species. With a common classification scheme, scientists could share knowledge and publish findings more easily.

Identifying and naming plants took a lot of work and understanding how they can produce food was even more difficult. Discoveries were gradual and came from many different sources.

Robert Hooke (1635–1703) had named **cells** in 1665. He coined that name because through his microscope the building blocks of plants resembled the chambers, or cells, in which monks worked and lived. Robert Brown (1773–1858) identified the **nucleus,** a central part of the cell, by 1831. By 1838, Matthias Jakob Schleiden (1804–1881) stated with confidence that all plant tissue is made of cells. But what was the job of the green cells that made up the leaves?

We now know that the leaves of plants are factories producing substances we can't live without. Plants create food and also release **oxygen,** a gas vital to our survival. Many scientists contributed to our learning about plant chemistry over time. In 1772, Joseph Priestly (1733–1804) proved that plants can make air livable again.

Priestly found different kinds of gases in air but could not identify for sure that it is oxygen that provides the breathable air. In 1789, Dutch scientist Jan Ingenhousz (1730–1799) found that without sunlight, plants will not **emit** the gas Priestley's plants produced.

Antoine Lavoisier's (1743–1794) discoveries about oxygen were essential to understanding how plants work. Lavoisier helped discover that burning and breathing are both cases of oxygen being consumed. Sadly, Lavoisier was executed during the French Revolution.

Gradually, people gave up the incorrect idea of **phlogiston,** supposedly a substance within matter that makes it burn. Instead, burning occurs when oxygen combines with the molecules of other matter.

Oxygen was found to be one of the different kinds of "air" separated by Priestley. Ingenhousz was able to figure out that plants have two respiratory (breathing, or exchanging gases) cycles and what the gases are. During the day, plants take in carbon dioxide ($CO_2$) and give out oxygen ($O_2$). In the night the process is reversed.

In 1837, Henri Dutrochet (1776–1847) demonstrated that parts of plants contained a green **pigmented** chemical called **chlorophyll.** Chlorophyll absorbs $CO_2$ when light is present. Julius von Sachs (1832–1897) discovered that chlorophyll is found in small bodies within the cells called **chloroplasts.** Von Sachs described the process taking place:

$$CO_2 + H_2O \longrightarrow starch + O_2$$

This formula is the basic equation for **photosynthesis.** Plants take in carbon dioxide and water and create **starch** and oxygen. The large food molecules produced by plants during photosynthesis are also known as **carbohydrates.** These carbohydrates can include **glucose** and other **sugars,** as well as starches. Plants also produce water vapor.

As you can observe in your experimenting, sunlight provides the energy for this chemical reaction. Plants are the original solar cells, using the energy of the sun to transform matter. Animals (such as humans) can eat the stored food in plants and breathe the oxygen they release.

## Vocabulary Words

botany .......................................... the study of plants

carbohydrates ......................... group of organic compounds composed of carbon, hydrogen, and oxygen atoms

cell ................................................ the smallest, microscopic-sized unit of organisms

chlorophyll ............................... green pigments that trap light for photosynthesis

chloroplast ............................... organelle containing chlorophyll that functions in the performance of photosynthesis

emit ............................................. to send out or discharge

glucose ...................................... common sugar composed of molecules of carbon, hydrogen, and oxygen atoms

nucleus ...................................... in a cell, the part separated by a membrane and containing most of the cell's DNA. In an atom, a positively charged mass containing protons and neutrons.

oxygen ....................................... gas that makes up about 21% of Earth's atmosphere.

phlogiston ................................ non-existent substance that early chemists mistakenly believed to be released in the form of flame when things burned

photosynthesis ...................... process by which plants convert carbon dioxide and water into carbohydrates using energy from the sun

pigmented ............................... colored

renaissance ............................. a revival; a reawakening of cultural achievement

starch ......................................... name for molecules of carbohydrates abundant in plants

sugar ........................................... general term for some large carbohydrates

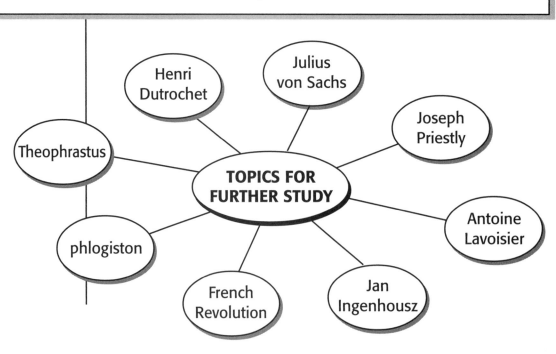

From *Science Giants: Life Science* © Good Year Books. This page may be reproduced for classroom use only by the actual purchaser of the book. www.goodyearbooks.com

# Propagating Plants

## Experimenting with Seeds

**Materials per Team**

- commercially packaged seeds (beans, peas, marigolds, etc.)
- potting soil
- planting trays, pots, or paper cups
- house or garden plants for cuttings or divisions

Students learn best about experimentation and controlling variables by testing their own hypotheses. Teach students the scientific method by *doing* the steps rather than by memorizing them. In this activity, encourage teams of students to devise their own experimental conditions. What do they want to find out? How can they set up a test and discover something about plant growth?

## Activity

To begin, open the pack of seeds and show that none of the seeds has sprouted. Ask students what the seeds will need in order to begin growing.

List students' suggestions and form groups to test the various claims. Possibilities will include sunlight, water, soil, and air. Remember to test combinations of key plant needs, such as air and water together.

Before students begin, emphasize these ideas:

▶ **Isolate variables:** Change only one thing at a time so you can see if that change makes a difference.

▶ **Establish controls:** Try your experiment without the important change to see what happens normally and to be able to make a comparison.

To begin, give teams the materials for the activity and have each team set up an experiment, testing one variable and maintaining a control. Use the chart on page 89 as a guide.

If a team hypothesizes that a seed in soil without water will not sprout, the tester must experiment with at least two sets of seeds. Have them tend one set with water, soil, and so on, and give another set the same treatment in every way *except* for water. Water is the variable being changed in the experiment. The set of seeds receiving water is the control group. Be sure they record their observations.

Encourage multiple sample sets in each test. The larger the population in each sample, the more accurate the data tend to be. The seed packet may even state what the expected germination rate of the seeds is based on the company's tests. Some seeds may not sprout no matter how they are treated.

Copy the chart on page 89 for each group or have students develop their own data recording sheets. Discuss each experiment before starting. For example, ask: How can you test seeds without soil? Brainstorm ideas together; you can suggest soaking in a cup, on a damp paper towel or sponge, sprayed each day and kept moist, and so on.

Be sure that all samples are subject to the same temperature and sunlight conditions as the planted seeds. Check each sample periodically, and repeat and refine the experiments based on the results. Use the other column for creative student ideas—refrigerating seeds, keeping them under a light all night, or other tests.

For a further look at botanical experimentation, encourage the students to devise tests for the sprouted plants and go on to the next activity ("What Plants Need"). The activity "Hitchhiking Seeds" demonstrates some of the ways nature disseminates seeds. Together, these activities will give students a sense of the cyclical life of plants.

Two other useful ways to demonstrate plant reproduction involve cloning. Taking cuttings from house or garden plants (e.g., coleus, geraniums, impatiens) and rooting them in water or potting soil shows the way many plants can make copies of themselves. Other species of plants produce "baby" versions of themselves that can be separated and potted to create new plants. Spider plants and some bromeliads may be used for separating new plant growth to make independent organisms.

Name _____ Date _____

# What Do Seeds Need?

Record how many of each sample you test. Put a check mark in each column that describes what you added to the seed. Remember to have a control sample set.

Count how many of your seeds sprouted after one week and how many sprouted after two weeks.

| Set # | Quantity Planted | Water | Air | Sunlight | Soil | Other | 7-day check | 14-day check |
|-------|------------------|-------|-----|----------|------|-------|-------------|--------------|
| 1 |  |  |  |  |  |  |  |  |
| 2 |  |  |  |  |  |  |  |  |
| 3 |  |  |  |  |  |  |  |  |
| 4 |  |  |  |  |  |  |  |  |
| 5 |  |  |  |  |  |  |  |  |

*READING:*

# New Plants from Old

Plants reproduce in a variety of ways. We think of most plants as growing flowers and producing seeds and the seeds growing into new plants. Flowers allow plants to produce **offspring** with a combination of properties from two parents. **Pollen** from one plant may land on the flower of another plant and unite with it to grow a seed. The seed carries information about how to grow and live from both parent plants.

Plant species evolve and change by **natural selection.** Positive qualities get passed along to the next generation as those carriers live to create new seeds. Negative qualities die out because the plants carrying them are not likely to survive and produce seeds. People can influence this process by combining the desirable traits of two plants together.

But many plants can grow clones of themselves. They do not need to exchange pollen with other plants. If you get a chance to propagate (start) a new plant from an existing plant, the new one will be a copy of the older one. All the **genetic** information it carries directing it to grow is inherited from its one parent.

Scientists have studied how plants reproduce for thousands of years. Farmers understood long ago that they needed to save seeds for next season's crops. When early **botanists** studied plants, they paid close attention to the flowers of each species. John Ray (1627–1705) was different. In 1660 he developed a classification system based on the types of leaves that emerged from the seeds. Ray's system became the basis for major divisions of flowering plants.

Exploration and trade brought new species to the attention of Europeans. For example, German **naturalist** Alexander von Humboldt (1769–1859) traveled to South America with Frenchman Aime Bonpland (1773–1858). Humboldt noted that plants growing at high altitudes near the equator resemble those near sea level in northern Europe. If you research Humboldt's achievements and travels, you will have a sense of the explosion of scientific knowledge in the natural sciences during his lifetime.

The diversity and abundance of plant life in the Western Hemisphere impressed European explorers, including Charles Darwin (1809–1882), who is famous for explaining the theory of evolution. Scientists are still discovering new species of plants in **rain forests** deep within South America.

Kew Gardens in London became a world-famous greenhouse center. Plants from warm, tropical climates could be grown in a northern European city. Joseph Banks (1743–1820) had traveled with noted explorer Captain James Cook (1728–1779) to the South Pacific and helped establish the plant collection at Kew Gardens.

Plants migrating from one ecosystem to another can cause a lot of trouble. At "home," a plant may be held in check by natural enemies. Animals eating it and diseases attacking it can hold down its population and prevent spreading and overcrowding. If it does spread to a new habitat, a plant's controls—such as animals who eat it and other plants that successfully compete for space—may be missing. Can you find any examples of **non-native plant species** that have invaded an ecosystem and become a threat to local plants?

Flowers produce pollen that can carry genetic information to other plants.

## Vocabulary Words

botanist .................................. plant scientist

genetic .................................... affecting or affected by genes; hereditary

natural selection ................... the survival of individuals whose characteristics are advantageous for their environment and elimination of those individuals who do not succeed

naturalist ................................ scientist who studies natural objects and organisms

non-native plant species ..... species that has been transported by human activity to a part of the world where it did not previously exist naturally

offspring ................................ descendants of organisms—for example, the children of human parents

pollen .................................... material produced by anthers of flowers that is the male element in fertilization

rain forest.............................. region of the Earth characterized by high annual precipitation and year-round green forest

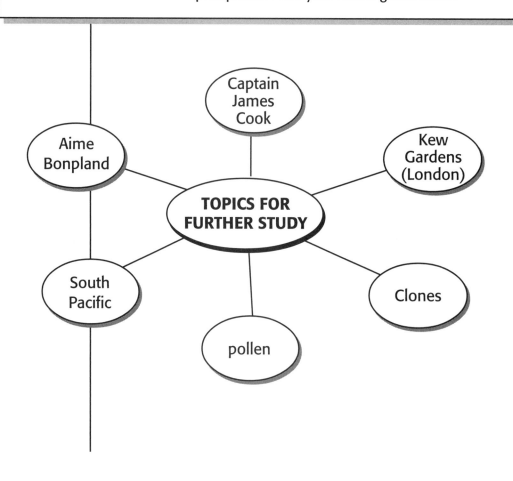

# What Plants Need

## Experimenting with Plants

Studying plants provides many opportunities for students to practice sound scientific methods while learning important content. Rather than reading about what plants need to thrive, students should be handling and observing plants. When students design their own experiments, there is a double learning payoff in content and process.

### Materials per Team

- sprouted plants, possibly from "Propagating Plants" activity
- variety of materials for setting up test conditions

---

## Activity

The experiments with seeds ("Propagating Plants" activity ) might have already provided specimens for a new set of tests. Be sure to use plants that sprouted from the control group so students can study a new set of variables without the results being affected by things they tried before. If your class does not have plants from earlier work, start a new batch of plants from seeds or obtain some plants.

Brainstorm a set of factors about plant growth students are interested in testing. Does commercial plant food speed the growth of plants? Does the temperature of water affect their growth? Use the list generated to discuss how we can find things out using the scientific method.

Help teams of students choose an experiment to design and implement. The steps of the scientific method are an integral part of the inquiry process, and students will become familiar with them by applying them. While students are planning their experiments, teach or review the scientific method. Ask:

1. What do you want to find out?
2. What is your prediction, or hypothesis?
3. What steps will you follow? How will you set up a control and test one variable at a time?
4. What is your data?
5. What does it tell you?

Here are some suggestions for experiments if you need to offer some guidance. Remember to compare the test plants to a set of control plants:

▶ Add plant food to water.

▶ Limit the hours of sunlight.

▶ Use a plant light to extend hours of light.

▶ Place the plant in the refrigerator during the night to vary temperature.

▶ Test the effect of "acid rain" by adding small amount of lemon juice to water.

▶ Compare the effects of direct sun to indirect sun.

After a few suggestions, students will come up with ideas of their own to test.

When each team has decided upon an experiment, be sure they have at least two similar plants and preferably more. Have them divide the plants into two groups, the control group and the experimental group. Students will change one variable in the experimental group and observe the plants for one to two weeks. Encourage daily observation and recording, and have them share conclusions with the whole class.

## READING:
# *What Do Plants Need?*

Plants are one of the major groups, or **kingdoms,** into which living things are divided. It is not as easy as we think to find properties that are unique to all plants. Several characteristics are common to most species, however. For example, most plants can produce their own food through **photosynthesis.** Most plants cannot move about on their own. Can you think of other common descriptions of plants?

Plants are complex organisms consisting of many types of **cells.** Individual cells have thick walls of a material named **cellulose.** When you study plants at home, in the field, or in the classroom, you will discover how different the parts of plants can be. Think of the plants we eat. Depending on the type of plant, we are consuming the **roots, stems, leaves, flowers,** or **seeds.** Can you think of examples of foods for each different part of plants?

If you grow a plant in a pot, you have to water it regularly. The water disappears and the plant grows. Naturally, people believed that plants grew by somehow turning the water into plant cells. In 1675, Marcello Malpighi (1628–1694), an Italian scientist, published a book about plant anatomy. Malpighi used a microscope to discover tiny details of plant structure. Nehemiah Grew (1641–1712) published a detailed plant anatomy book in 1682. But the secrets of plant **physiology** were to remain hidden for many more years.

Stephen Hales (1677–1761) conducted careful experiments on plants. He changed lots of **misconceptions** that were accepted at the time. In the 1720s and 30s, Hales measured the water and soil along with the mass of growing plants and learned that plants do not circulate liquids in the way animals circulate blood. He reported that liquids travel through the plant and gases exit through the leaves. Hales also learned that a part of the air is necessary for plant **nutrition.**

Meanwhile, Swedish naturalist Carolus Linnaeus (1707–1778) was classifying and studying up to eighteen thousand plant species. But how were plants producing so much growth? Animals eat plants or other animals for food, but plants seemed to grow their own food out of the air!

That is exactly what Nicholas de Saussure proved with his experiments in 1804. He found that plants need **carbon dioxide** ($CO_2$) from the air in order to grow. De Saussure found that nitrogen from the soil is also necessary, but most of a plant's mass comes from the carbon in the air. By 1840, Justus von Liebig (1803–1873) could explain the carbon and **nitrogen** processes of plants and animals.

Julius von Sachs (1832–1897) was an early experimenter in hydroponics, or the growing of plants in water. He helped discover that plants take carbon from the air for photosynthesis and need traces (small amounts) of other **elements** also. These micronutrients include potassium, calcium, nitrogen, phosphorus, magnesium, and sulfur. If your family maintains a lawn and buys fertilizer for it, look at the bag and see what chemicals are included. Read the analysis of chemicals in indoor plant food also, and you will see the elements von Sachs identified.

Farmers had known for years that rotating, or moving, the position of crops maintained the richness of their fields. Nineteenth-century scientists discovered why—the nitrogen (and other chemicals) in the soil were used by plants and needed to be refreshed and replenished. In order to use $CO_2$ to produce food, plants had to have a supply of nitrogen.

When you do your experiments to learn what plants need, remember that it is not always easy to see where they get their necessary supplies. **Botany** continues to use the knowledge of **chemistry** to learn more about the life and behavior of plants. Advancement in science depends on an interconnected web of learning.

## Vocabulary Words

| | |
|---|---|
| botany | the study of plants |
| carbon dioxide ($CO_2$) | common gas composed of molecules with one carbon atom and two oxygen atoms |
| cell | the smallest, microscopic-sized unit of organisms |
| cellulose | carbohydrate used by plants as a structural material |
| chemistry | science of the structure and properties of matter |
| element | substance composed of one type of atom |

## Vocabulary Words *(continued)*

flower ........................................ reproductive structure of a seed-producing plant; blossom

kingdom .................................... largest (most inclusive) category used in common method of classifying organisms

leaves ........................................ green structures of plants attached to stems, functioning primarily in photosynthesis

misconception ........................ a mistake in meaning; the state of having the wrong idea about something

nitrogen .................................... gas that makes up almost 80% of the atmosphere

nutrition .................................... process or study of how organisms use food

photosynthesis ........................ process by which plants convert carbon dioxide and water into carbohydrates using energy from the sun

physiology ................................ study of vital life processes

roots .......................................... part of a plant that anchors it, helps absorption, and may store food

seed .......................................... fertilized ovule of a plant, capable of growing into a new plant under favorable conditions

stem .......................................... stalk, slender connecting part of a plant supporting a leaf or flower

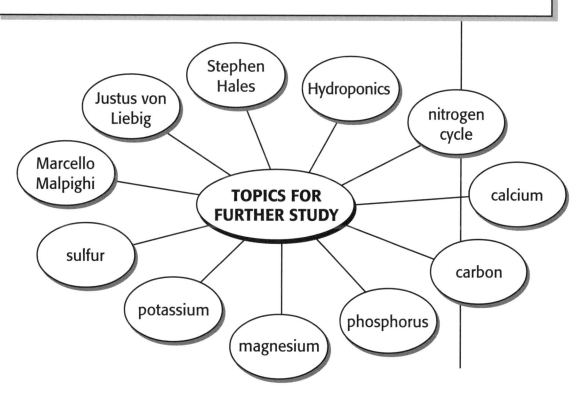

## Materials per Team

- old, clean socks
- potting soil
- clear containers with lids (e.g., terrarium containers) or containers with plastic-wrap covers
- access to a weedy area

# Seeds and How They Spread

## *Hitchhiking Seeds*

Seeds are the primary vehicles of reproduction for most plants. But how do plants spread over new territory? This activity, which is best done in the fall, demonstrates one method.

### Before the Day of the Activity

1. Ask students to donate old but clean socks (the larger the better) for science class. Why there are so many single socks missing a mate seems to be an unsolved mystery of laundry science, but that's a problem for another day! At least we can provide a use for the leftover single socks.

2. Find some likely terrain where weeds might flourish. Travel through a meadow or edge of a woodland, the border of a schoolyard too close to a fence to be mowed regularly, or a clearing where plants are allowed to grow naturally. Vacant, untended areas are the best places to find locally dominant weeds proliferating. Be sure to preview the area. Look for poison ivy or other allergy-provoking species, and scout out points of interest and valued plants to protect from trampling.

### On the Day of the Activity

On the day of the activity, tell students that they will be going on a nature walk. Then:

1. Give each group a clear container with a lid or provide containers with plastic wrap to be used for lids. These will be their terrariums. Have students partially fill the terrariums with potting soil that should be free of seeds.

2. Next, have students slip the donated socks over their shoes.

3. Break the class into small groups. This will ensure that an area does not get overly trampled and that walkers will brush up against a variety of plants.

## On the Walk

As they walk, remind students to show respect for nature by treading lightly. As students walk through weedy areas, note which plants have bloomed and set seed.

## After the Walk

After the walk, have students remove the socks and either shake or comb them out over the potting soil in their terrariums. They can also put the socks directly in the containers. They should then add water and put on the cover. Tell students to observe these terrariums over the next week or two, continuing to water them regularly and compare the terrarium plants to the ones they observed on the walk.

Ask: Which plant seeds "hitchhiked" indoors on the socks? What animals might serve as vehicles for the hitchhikers in the wild, even though they don't wear socks? Some varieties of seeds have interesting parts that hook onto passing creatures. These make fascinating viewing under lenses and low-power microscopes.

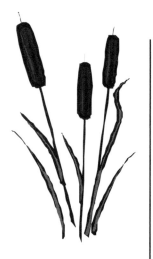

# READING:
# *Traveling Plants*

It's pretty easy to figure out how animals can move from place to place. Some walk, some gallop, some crawl, some fly, some swim, . . . you get the idea. But most plants are rooted in place. How can they spread out around the world?

Since Galileo's time and even before, scientists have learned by careful observation and by experimenting. Where can they find places to study the spread of plants? Many places in the world are already covered with plants, but once in a while nature creates a laboratory where species have to start over. Volcanoes sometimes clear out all vegetation and provide a clean slate for plants.

In the late nineteenth century, one of the greatest eruptions of all time demolished the island of Krakatau in Indonesia. The effects were felt around much of the world. Essentially, brand-new land was created. Scientists have been studying the colonization process by plants there for more than one hundred years. This process is called **succession,** or sometimes *plant* or *forest succession.* Pioneer plants first cover the disturbed land. Over time, a series of plant species dominate the land, leading to a stable community. Other opportunities to observe plant succession occur when nature or people disturb the existing ecosystem. In 1980, Mount Saint Helens erupted with great force in Washington State. Scientists found a learning laboratory in the Pacific Northwest where succession can be studied. In 2004, Mount Saint Helens became active again.

Human action can also erase the work of nature and offer a glimpse of how plants invade a new territory. Farmers clear fields to grow the crops they choose. When they stop working their fields, farmers allow nature to take its course to erase the work of people. In New England, many areas of forest have grown back, allowing scientists to study the progression of grasses, bushes, and trees.

In many parts of the world, fires have cleared areas, opening opportunities for **pioneering species.** Approximately 45% of Yellowstone National Park burned in 1988. Fire is a natural factor in many ecosystems and **fire management** is a complicated and **controversial** issue.

The study of organisms and habitats together is called **ecology.**
Charles Darwin (1809–1882), most famous for developing the
theory of evolution, was keenly aware of the importance of this
connection. Organisms are naturally selected, he **hypothesized,**
based on their success in adapting to the factors in their environment.
When an area is suddenly cleared, scientists can watch the
competition among plants and animals to establish a community
while the playing field has been opened.

## Vocabulary Words

| | |
|---|---|
| controversial | describes an issue or idea that causes a dispute |
| ecology | study of the relationships among organisms and the environment |
| fire management | technique in which controlled fires are allowed to burn to maintain the health of an ecosystem |
| hypothesis | explanation or theory |
| pioneering species | hardy organisms that first colonize an area after a drastic change |
| succession (plant) | orderly sequence of change of species occupying an environment over time |

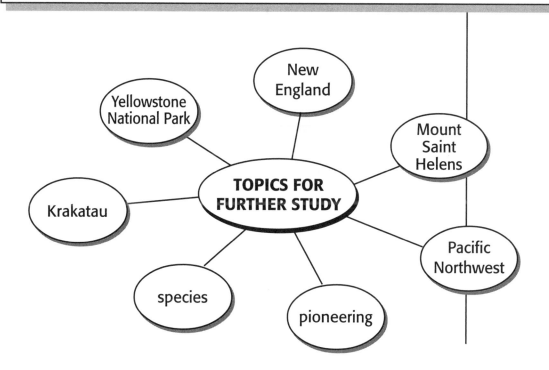

New England

Yellowstone National Park

Mount Saint Helens

TOPICS FOR FURTHER STUDY

Krakatau

Pacific Northwest

species

pioneering

## Materials per Team

- large flowers for study and sketching
- "deadheads" for dissection
- magnifying lenses
- reference texts with flower drawings (encyclopedias, plant identification books, science texts, etc.)
- paper and pencils

# Parts of a Flower

## *Flower Dissection*

Learning science sometimes seems like learning a foreign language. Students who are strong scientific thinkers but who have difficulty with language skills may be held back by overwhelming vocabulary. In addition, students may become "turned off" to science if they struggle with definitions and lose sight of the discovery and inquiry nature of scientific activity. This is a dilemma for teachers.

This activity teaches scientific vocabulary using hands-on experiences. Students will have a chance to study the structure of flowers and learn the names of their parts.

Your dilemma is to figure out how best to teach this material. You know your students and their learning styles best. Is it better to teach them the flower names and then use the actual materials, or should the labeling follow the investigation? Perhaps a simultaneous approach is best: While students are viewing and drawing the flowers, provide reference texts with the names so they can fill in their diagrams.

### Before the Activity

Before the day of the activity, obtain some large flowers. Use cut flowers from a florist or flowers taken with permission from a garden, or head outside for field study. If you have access to different flower species, use them for variety.

If there are a lot of gardens in the area, you might find that "deadheads" will be available. Many gardeners snip off faded flower stalks to discourage seed production in plants grown from

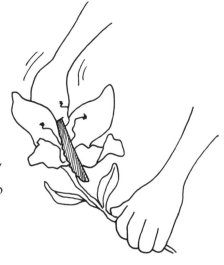

bulbs. Daffodils and even some tulips have an improved chance at growing strong flowers next spring if their energy is used to build leaves and roots rather than seeds. Obtain as many cut flower stems as you can so each student can dissect his or her own flower.

## On the Day of the Activity

Set up stations, one with each flower. Allow students to circulate through the stations, or have each team study one flower. They should use the following steps in the inquiry model as they move through the activity:

- ▶ observation and hands-on investigation
- ▶ information processing
- ▶ sharing
- ▶ consolidation/assessment

1. First, tell students that in this activity they will be observing and drawing plants. Remind them that plants have a cyclical lifestyle. When flowers fade, the seed production process takes much of a plant's energy. Whether students are observing flowers indoors in water or outdoors in soil, the petals will fall and they can then study the inner parts.

2. Encourage students to use magnifying lenses and microscopes without slides to observe the exterior flower parts. Dissecting microscopes reflect light off the object, not through it, and often present a beautiful three-dimensional view.

3. Provide reference texts with botanical drawings so that students can begin naming the flower structures.

4. Have students sketch the external flower parts as they observe and investigate them (see the drawing on page 104). They can consult their reference materials as they sketch. Remind them to label the parts of their flowers in their sketches.

5. Hold a class discussion. Ask students to share their drawings and talk about the structures they found in their flowers. Compare findings between groups.

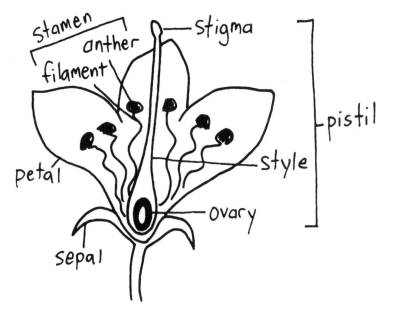

Now students are ready to investigate the interior flower parts. Use the same procedure for the interior as you did for the exterior parts. In general, no sharper tool than a fingernail is needed to slice open the chamber below the flower and expose the parts, but students can also use plastic table knives. Encourage students to cut through the pistil down toward the stem far enough to see the ovules, the beginnings of next year's seeds.

Students should begin to gain an understanding of how and where seeds form. If you do the activity in the spring, they may be motivated to watch the progression of other flowering plants with a more interested, informed eye.

# READING:
# Flower Power

When people and most other animals are born, they inherit characteristics from both a mother and a father. Traits that may not show up in one generation can be carried along. Information telling cells how to grow, work, reproduce, and even die is coded through the language of **DNA.** Because animals mix their DNA with that of a mate when they produce **offspring,** species change. They can adapt to **environments** over time. If **mutations** occur, **natural selection** can act to either keep the change in the DNA code when the animal is successful and reproduces or eliminate it when the animal dies without producing offspring.

How about plants? Do they have a similar mechanism for mixing their genes? In 1580, Prospero Alpini (1553–1616) of Italy began to investigate this question. Although he didn't know about **genes** and DNA, Alpini wrote that plants reproduced with male and female parts. Some plants have their male and female parts in the same flower while other species separate them. Carolus Linnaeus (1707–1778), the famous Swedish **classification** specialist, emphasized flower structure when he named and sorted plant species. In 1877, Charles Darwin (1809–1882) noted that some species of plant grew different forms of flowers on different plants.

Just as plants have evolved different ways to spread their seeds, they have also evolved different ways to mix the genetic information in those seeds. Insects and other creatures are attracted to flowers for a **nectar** meal and in the process deliver **pollen** from flower to flower. The wind can spread pollen, too, as many **allergy** sufferers know. Pollen can carry genetic information from one plant to be shared in seed formation on another plant.

Plants are very resourceful. There are few places on Earth without *any* of these vital organisms. Plants are the beginning of most food chains and keep much of the rest of the world fed and breathing. And they continue to **evolve,** along with the other **kingdoms** of life. If you study **endangered species,** don't overlook the importance of plants to the rest of the community of life on Earth. Many plant species are threatened or endangered.

## Vocabulary Words

| | |
|---|---|
| allergy | physical reaction to a substance in the environment |
| classification | system of grouping organisms based on their similar characteristics |
| DNA | deoxyribonucleic acid; a double twisted helix inside a cell that contains the genetic code for that organism |
| endangered species | species in danger of becoming extinct |
| environment | an organism's surroundings |
| evolve | act of changing gradually |
| genes | sections of DNA that produce a trait in an organism |
| kingdom | largest (most inclusive) category used in common method of classifying organisms |
| mutation | permanent change in the genes of an organism |
| natural selection | the survival of individuals whose characteristics are advantageous for their environment and elimination of those individuals who do not succeed |
| nectar | sweet liquid secreted by flowers |
| offspring | descendants of organisms—for example, the children of human parents |
| pollen | material produced by anthers of flowers that is the male element in fertilization |

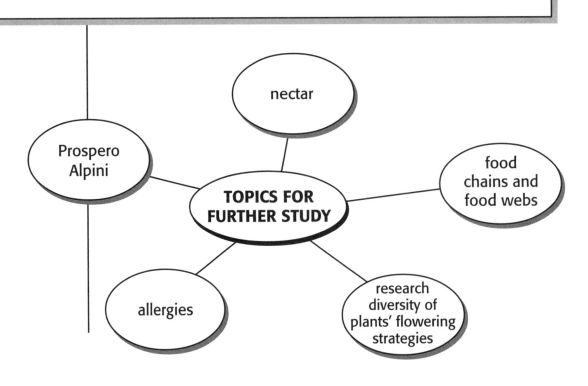

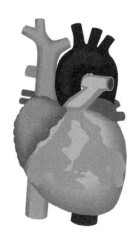

# CHAPTER 5
# Human Body

**TIME LINE**

| Year | Notable Event |
| --- | --- |
| 400s B.C. | Greek scientists studied the heart and blood, realizing their importance to life. |
| A.D. 161 | Galen practiced medicine in Rome, establishing practices that will be followed for centuries. |
| 1242 | Ibn al Nafir wrote about the pumping action of the heart |
| 1543 | Andreas Vesalius wrote an anatomy book. |
| 1616 | William Harvey lectured about blood circulation to students in England. |
| 1665 | Marcello Malpighi described the nervous system as fibers connecting to the spinal cord and then to the brain. |
| 1669 | Richard Lower noted that blood changed color depending on its place in the body. |
| 1796 | Edward Jenner performed a vaccination to protect a boy against smallpox. |
| 1846 | William Morton used anesthesia during operations. |
| 1861 | Pierre-Paul Broca demonstrated that specific parts of the brain govern certain functions. |
| 1865 | Joseph Lister used carbolic acid to disinfect wounds. |
| 1880 | Louis Pasteur publishes his work on the germ theory of disease. |
| 1895 | Wilhelm Roentgen discovered X rays |
| 1957 | Christiaan Barnard performed a heart transplant. |
| 1968 | Denton Cooley implanted an artificial heart in a human patient. |

## Materials per Team

**Part 1**
- modeling clay
- toothpicks
- skeleton posters

**Part 2**
- cleaned bones left from a meal
- disposable gloves (optional)

**Extension**
- owl pellets

# The Skeleton

## *Model Skeletons*

Humans and other vertebrates can grow to a large size and achieve a wide range of movement. An internal skeleton allows for many body parts far beyond the range of invertebrate animals. The following activities will encourage students to investigate the nature of bones in humans and other creatures.

## Part 1

Children have a lot of prior knowledge about bones. Building a model is a good way to test their assumptions. As they construct skeletons out of clay and toothpicks, students will reveal what they know, and what they want to find out.

Try to obtain a poster of a human skeleton and use other media (books, Internet, etc.) for student reference. Have a short discussion before building the skeletons. Suggest that the toothpicks represent bones and the clay represents places where the bones meet.

Toothpicks can be broken to make different-sized bones. Investigate some of the types of joints as a class. Compare the range of motion at the shoulder compared to at the elbow. Ask what the connection of bones at the spine is like. Decide whether to have groups or individuals build the skeletons.

When students are done, compare the models. Besides learning anatomy, children are also tackling a design and engineering problem when they build a model.

## Part 2

After sharing the learning gained from model building, bring in some chicken and fish bones or even a whole fish for dissection. Looking at real bones after making a model is a good follow-up.

Ask students to bring in bones from meals to share in class. A bath in a weak bleach solution will clean the bones pretty well, but be sure everyone washes carefully after handling animal parts. Using inexpensive disposable gloves is not only hygienic but adds a scientific atmosphere to the activity and may coax reluctant students to become hands-on investigators.

## Extension

Many teachers purchase owl pellets from science supply companies to use with classes. The contents of an owl pellet are dry and consist of the indigestible parts of the owl's prey. That may sound unappealing, but the pellets are sanitized and the activity is highly engaging.

Students separate fur from bones while sorting the bits and are usually able to find and assemble several mammal skeletons per pellet. The only tools required for dissecting owl pellets are tweezers, plastic knives, and straightened paper clips. As students compare rodent skeletons to human skeletons, encourage them to list similarities and differences. They will be surprised to find many shared characteristics among mammals, despite the huge size difference.

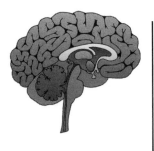

*READING:*
# Looking Inside the Body

Anatomy, the study of the structure of the body, is one of the oldest sciences. Caring for the ill required a familiarity with body systems. In ancient Greece, Pythagoras (c. 569–475 B.C.) performed **dissections** on dead bodies to increase his knowledge of **anatomy.** In this way, Pythagoras discovered the function of the brain, among other findings.

Hippocrates (c. 460–376 B.C.) was the most famous doctor of ancient times. He studied diseases scientifically rather than magically. Modern doctors are still taught the Hippocratic Oath, advising them first and foremost to do no harm to patients.

In the **Middle Ages,** a person might have a job that combined the tasks of a barber and a surgeon. Most medical practice dated back to the ideas of Galen (c. 130–200), a Roman physician who had written a major textbook about medicine. It was not until the sixteenth century that European scientists returned to investigating human specimens openly. Medical science became an important specialty. Better and more useful tools were developed to improve diagnosis and treatment of disease.

Andreas Vesalius's (1514–1564) anatomy textbook written in 1543 set a modern standard of knowledge. The illustrations were based on actual dissections, not ancient folklore. When the microscope was developed in the 1600s, the tiny became visible and a new world opened up to scientists. The body was made of connected systems, complicated yet able to be studied.

Most bones are large. It was fairly easy for students of Vesalius's anatomy study to understand the place of each bone in the body. Bones protect our internal organs. Think of the skull as a helmet for the brain and the ribs as a shield for our heart and lungs.

But bones have other roles to play besides that of a framework for holding up the rest of the body. Bones also produce **red blood cells.** This process takes place in the spongy interior layer containing **marrow.** A different kind of marrow at the core of bones is one of the body's storehouses for fat.

Some bones can move at the **joints** where they meet other bones and some bones are fixed in one place. In 1895, doctors gained the valuable ability to look inside the body and see bones.

Wilhelm Roentgen (1845–1923) discovered how to direct what he called **X rays** toward and through a patient's body, revealing the bones within the body.

As you learn more about the structure of bones and the many jobs they perform, you can understand why it is vital to keep them strong. The chemistry of bones is complicated, but calcium is a major building block. Sometimes when people age, their bodies lose calcium and their bones may become brittle. That's why milk is one good way to keep your skeleton happy.

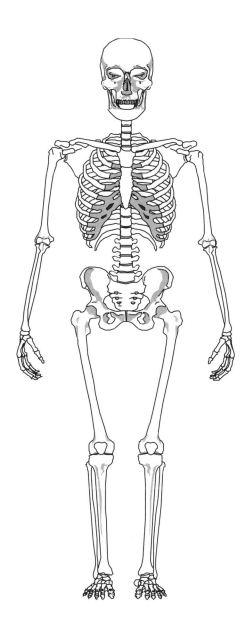

## Vocabulary Words

anatomy ...................................... the structure of an organism

dissection ................................. to cut apart for study

joints ............................................ connection between bones or parts of animal bodies that permit movement

marrow ...................................... soft tissue inside bone

Middle Ages ........................... historical period in Europe, approximately the years A.D. 500–1500

red blood cell ......................... a cell that transports oxygen in the body

X rays ...................................... radiation of a specific high-frequency wavelength in which the photons have high penetrating power and can pass through solid objects

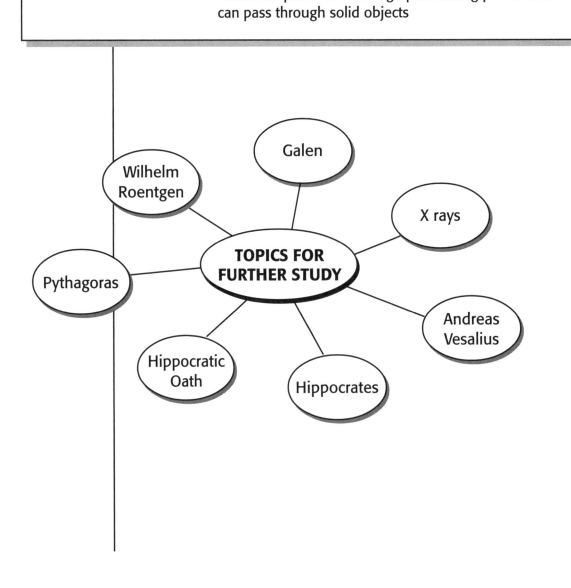

# Circulation

## Circulation Dance

From *Science Giants: Life Science* © Good Year Books. This page may be reproduced for classroom use only by the actual purchaser of the book. www.goodyearbooks.com

**T**his activity simulates the human circulatory system as students pretend to be the parts of the body. About half of the students portray blood cells and do the circulation dance. The rest of the students act out the parts of organs and other cells. The roles include the four chambers of the heart, two lungs, two kidneys, the intestines, and some generic cells. The number of other body parts you choose to represent with students will depend upon how many players you have.

### Materials per Team

- lively music
- four sets of cards
- signs for players
- colored paper

---

## Activity

To begin, create four sets of cards such as those shown below. Decorate and label them *oxygen, nutrients, carbon dioxide,* and *waste.* On the back of the nutrient cards, print ENER. On the back of the oxygen cards, print GY. The blood cells carry all cards around to the other parts of the body. Make at least two of each type of card for each "blood cell."

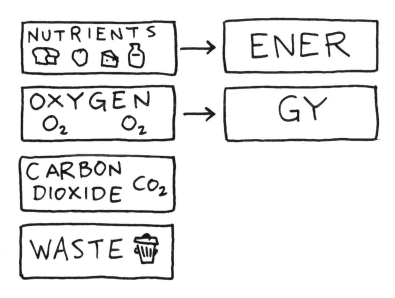

Set up the students in the pattern shown on page 115. Explain that while the music plays, the "blood cells" will dance through the circulatory system and carry products to and from the parts of the body.

Assign two students to represent each heart chamber. They can join hands around each cell and gently pump it a few times as it passes through. Blood cells must first enter the atrium and then the ventricle when passing through each side of the heart.

▶ The *lungs* begin with oxygen cards and give them to the passing blood cells in exchange for carbon dioxide cards.

▶ The *intestines* begin with nutrient cards to give the blood cells.

▶ The *body cells* (a generic representation of all the diverse parts of the body) start the game with waste and carbon dioxide cards and trade them to the blood cells for nutrient and oxygen cards.

▶ The *kidneys* receive the waste cards.

The dance continues until the music stops or the oxygen and nutrient cards are all distributed to the body cells.

# Diagram of Circulation Flow

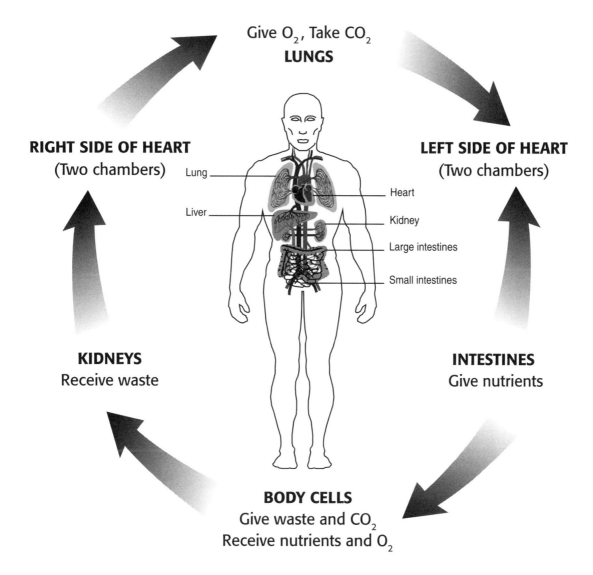

Give O$_2$, Take CO$_2$
**LUNGS**

**RIGHT SIDE OF HEART**
(Two chambers)

**LEFT SIDE OF HEART**
(Two chambers)

Lung

Liver

Heart

Kidney

Large intestines

Small intestines

**KIDNEYS**
Receive waste

**INTESTINES**
Give nutrients

**BODY CELLS**
Give waste and CO$_2$
Receive nutrients and O$_2$

When the cards are all at their final destination, the body cells can turn their cards over and see that oxygen plus nutrients spells ENERGY for them. Help students see that to work harder, the body needs more energy. That would mean faster circulation or more efficient delivery. Breathing hard during exercise, increased heart rate, hunger after a lot of work—all of these are signs that our body systems have been working hard, using oxygen and nutrients. Students may want to play again, switching roles so be sure everyone understands the job of each part of the body in this game.

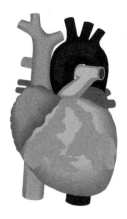

### READING:
# Keeping the Blood Flowing

As far back as we can trace, the heart has been recognized as one of the most important organs in the body. This sounds obvious to someone in modern times, but it took many years of clever thinking and research to understand its role in the body.

Thanks to the teachings of Empedocles (c. 492–432 B.C.), we know that the ancient Greek scientists understood that the heart somehow controlled blood. Hippocrates (c. 460–376 B.C.), who is famous as one of the first medical scientists, identified blood as essential to life.

Herophilus (c. 335–280 B.C.) and other Greek scientists performed **dissections** and actually examined the heart and other organs. Praxagoras (c. 340–280 B.C.) made an important discovery around 300 B.C. He wrote that there were two different types of blood vessels, **arteries** and **veins.** It was evident that the veins carried blood. But after a person died, the arteries were empty because the heart had stopped pumping. Praxagoras reasoned incorrectly that the arteries carried air.

By 290 B.C., Erasistratus (c. 304–250 B.C.) determined that every organ had both an artery and a vein connected to it. So the Greeks knew a lot about blood and the heart but did not understand circulation for many years. That breakthrough was led by Galen (A.D. 130–200), a Greek scientist living in what had become the Roman Empire.

Galen established that both arteries and veins carried blood. He determined that the arteries carried blood away from the heart and the veins returned it. But he missed an important fact. He knew the heart had four chambers, but he thought it worked as a single pump. This set of assumptions about blood circulation lasted for many centuries. Dissections were forbidden and most medical research stopped.

In 1242 a scientist in Syria named Ibn al Nafis (1210–1288) wrote about a major discovery: He observed that the heart pumped blood to the lungs and then on to the rest of the body. The heart is actually a double pump, with each half sending blood on different sections of its route.

Communication among scientists was poor in the thirteenth century. Nafis's discovery was not widely shared with the rest of the world. When a Spanish scientist, Michael Servetus (1511–1553), made the same discovery, his work also failed to spread. (Servetus was executed for favoring political views that conflicted with the leaders of his day.) An **anatomy** book written by Italian scientist Mondino de Luzzi (1275–1326) contained the old incorrect information about circulation.

Gradually the facts won out as the sixteenth century brought a fresher climate for scientific discovery. In 1543 Andreas Vesalius (1514–1564) updated medical science with an anatomy book based on direct observations and experimentation. Realdo Colombo (1516–1559) wrote about the circulation between the heart and lungs, and others accepted his ideas. Hieronymus Fabricius (1537–1619) discovered valves in veins. Fabricius explained that the valves prevented blood flowing back in the wrong direction.

William Harvey (1578–1657) was born in England but studied with Fabricius. Harvey found valves in the heart. He reasoned that blood circulated around the body and back to the heart. Many people thought that the body had to make blood continuously as it would be "used up" on its journey. In 1628, he published a book describing the circulation of blood and it led to many other advances in the study of circulation.

The use of microscopes by Marcello Malpighi (1628–1694), Anton van Leeuwenhoek (1632–1723), and others led to the discovery of the blood cells and their primary importance to life. Malpighi saw and named the tiny blood vessels that transport blood to all the parts of the body **capillaries.** English scientist Stephen Hales (1677–1761) reported measuring **blood pressure** in 1733.

Once circulation, blood cells, and blood pressure were better understood, the pace of discovery increased. But another interesting mystery remained to be solved. Richard Lower (1631–1691), an English scientist, wrote in 1669 about the colors of blood. Generally, blood in the arteries and outside the body appears red. Blood in the veins looks blue. Lower experimented with **transfusions** and **intravenous** feeding. At about the same time, chemists were unlocking the secrets of atoms and molecules. It is the oxygen in the air and the iron in the blood that combine to create a red color.

**White blood cells** were found to be disease hunters that could absorb and destroy foreign bodies in the blood. The work of Casimir-Joseph Davaine (1812–1882) and Ilya Mechnikov (1845–1916) helped identify the body's self-defense system. Soon, scientists discovered other chemicals carried in the blood. William Bayliss (1860–1924) and Ernest Starling (1866–1927) studied **hormones,** controllers of body growth and function. Other scientists learned about **blood clotting,** the level of **glucose,** and other systems vital to life that depend on the circulatory system.

In 1953, after many years of experimenting, a heart and lung machine built by John Gibbon (1903–1973) kept a patient alive during heart surgery. By 1967, Christiaan Barnard (1922–2001) of South Africa performed a heart transplant. In 1969, Denton Cooley (b. 1920) placed the first artificial heart in a human being.

One blood test gives a doctor lots of information about a patient. Scientists continue to learn more about the circulatory system, especially on a microscopic level. As you find out about these famous discoveries of the past, imagine what the discoveries of the future will tell.

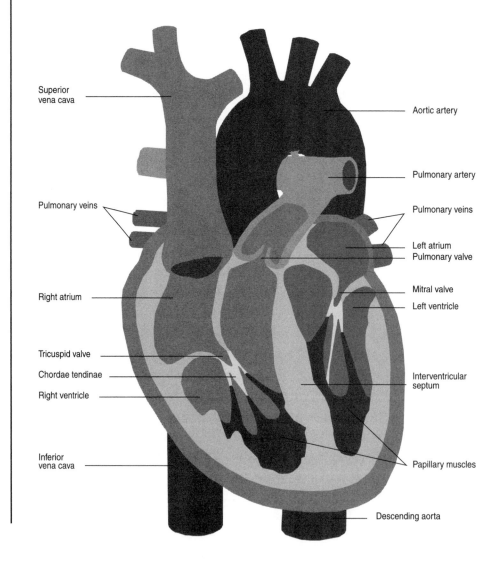

# Vocabulary Words

anatomy ..................................... the structure of an organism

artery ........................................ blood vessel carrying blood away from the heart

blood clotting ......................... massing of blood cells, designed to stop loss of blood from a wound

blood pressure ....................... force of the blood within the arteries

capillary .................................. smallest blood vessels, carrying blood between arteries and veins

dissection ............................... to cut apart for study

glucose .................................... common sugar composed of molecules of carbon, hydrogen, and oxygen atoms

hormone .................................. chemical produced by an organism that causes an effect on another part of its body

intravenous ............................ in a vein

red blood cell ......................... a cell that transports oxygen in the body

transfusion ............................. injection of blood or plasma into the bloodstream

vein .......................................... in animals, a blood vessel carrying blood toward the heart; in plants, a vascular structure forming the framework of a leaf

white blood cells ................... blood cells that fight and remove foreign bodies in the blood

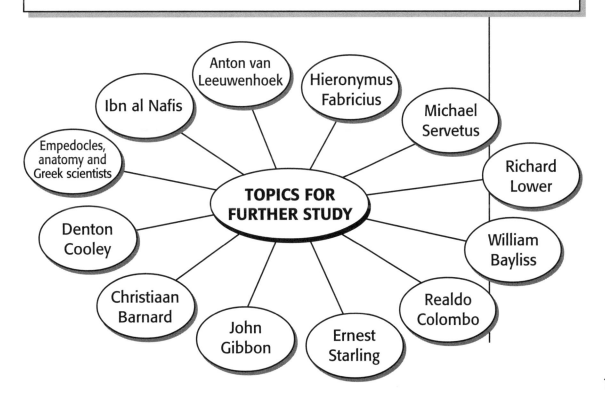

119

## Materials per Team

- cardboard tubes from paper rolls
- sets of same-sized objects (e.g., rulers, pencils, and erasers)

# The Senses (Eyes and Ears)

## *Two Eyes and Two Ears*

We have depth perception, or the visual ability to detect differences in distance, because our two eyes work together. Without this type of vision, called *binocular,* is depth perception possible? Students can work in pairs to test their depth perception while using only one eye at a time.

### Activity 1

To begin, have one child look along a desk at eye level and then close his or her eyes. Then have a partner place two identical-sized objects at different distances. The viewer opens one eye at a time and tries to determine which object is nearer and which is farther away. Be sure to have students vary the distances, objects used, and of course, which eye is being tested. Does using two eyes at once simplify the task?

An optical illusion creates another example of our eyes working as a team. Tell students to look through a cardboard tube with one eye while keeping both eyes open. Then they must hold a hand in front of the other eye. They should move the hand slowly away from the eye until they have the illusion of peering through a hole in the hand.

Our eyes work as a team and our brain combines the input coming from each one into a single image. The tube and the hand are joined into one picture, giving each of us the illusion that we have a hole in our palm.

## Activity 2

In a similar way as our eyes do, our ears work as a pair to help us judge distance. Here's a game to illustrate that skill. Have one player stand in the center of the room with his or her eyes closed. Point to another student who will then clap. The person in the center tries to guess where the noise came from. Repeat a few times with one person clapping at a time. Then repeat the game with the person guessing covering one ear. It's a lot harder to judge the direction of sound using only one ear.

*READING:*

# Human Senses Compared to Those of Other Creatures

If you were to draw a map of your neighborhood, you would likely place your home near the center. People have considered themselves the center of things throughout history. Next to gods and religious powers, humans have held themselves in an exalted position in the order of things.

The history of science reveals a tendency toward picturing our species in a more modest light. Rather than being the hub of the universe, our home planet Earth turns out to be a relatively small sphere orbiting a more or less medium-sized star. Our genetic material resembles "lower" animals in many ways. Ninety percent or more of our **DNA code** is identical to that of certain **primate species.** Humans are animals and can be classified and assigned a genus and species. Looking back only a short way into Earth's long history reveals a world without humans at all.

Perhaps the most amazing attribute about humans is how much we know, especially about ourselves. The pace of advancement in medical technology is truly astounding. Procedures performed routinely in the late twentieth and early twenty-first centuries were impossible mere decades earlier. What makes humans so inventive and successful?

Our senses are fairly modest in ability. Compare them to other animals.

Sight? Our eyes cannot see tiny objects from high above the Earth as hunting birds can, or in very low light like owls.

Hearing? Our ears do not detect sounds as effectively as the bats who bounce high-frequency waves off their prey and other objects.

Smell? We would do poorly tracking with our noses, a skill at which many other animals excel.

Taste? Compare our tongues to the **Jacobsen's organ** of a snake, a body part that reads signals deposited upon it by the snake's tongue. The snake seems to taste the air!

Touch? The speed of our sense of touch in transmitting signals to our brain and our muscles is impressive, but we lack sensitivity in hairs that some mammals have, such as a cat's whiskers.

How do humans do so well interacting with our environment with such modest equipment? First of all, consider the command center—our brain. It is an incredibly powerful organ, unequalled as far as we know among other organisms. Find out some facts about the brain and you will be impressed, too.

Second, once we receive information via our senses and our brain decides what to do, we are well equipped to build tools. Our hands have ten nimble fingers, including two very useful thumbs, and humans have built machines that bring our senses up to or close to the level of our fellow animals. For example, binoculars improve our distance vision, and sonar enables us to detect objects in a way similar to bats.

Third, modest or not, taken together, our senses are a well-rounded set of tools. Alone, each sense may be topped in performance by some other species, but together all of our senses, guided by a powerful brain, seems pretty good.

Let's consider some of the attributes of human sense compared to other species.

▶ **Sight:** Many **arthropods** (spiders, insects, crustaceans, and other animals) have **compound eyes.** Compound eyes are very sensitive to movement but may be limited in color range. Human eyes contain specialized receptor cells called **rods** and **cones** on the **retina,** the part of the eye that receives light signals. Rods help us see in low light, and cones help us see color and details.

Cats have better night vision than humans. Cats have a layer within their eyes called the **tapetum** that reflects light back so the rods have another chance to absorb it.

▶ **Hearing:** Human ears contain three small bones that transmit vibrations. Vibrations are transmitted through the air as waves with specific **frequencies** measured in cycles per second. The measure "cycles per second" is called hertz (Hz) in honor of Heinrich Hertz (1857–1894), a pioneer in radio wave discovery. Humans can hear sounds from about 20 to 20,000 Hz. How does that compare to other animals? Most fish and reptiles do not hear sounds in the higher ranges, while there are dog whistles that are higher than people can hear. Scientists have researched and charted the hearing range for different creatures.

▶ **Smell and Taste:** These senses are linked. Both depend on **chemoreceptors,** nerve cells that react to chemical stimulation. Humans have **taste buds,** receptor sites on our tongues that react to substances in our mouths. A popular theory says there are four basic tastes, each with a part of the tongue especially sensitive to it. But we actually sense a lot more than sweet, sour, salt, and bitter tastes. Our nose helps us distinguish among different foods, and combinations are practically endless. Think about how primary colors can combine to create countless different variations and you can imagine how taste types can join to add complexity to our meals. There are also lots of sours and lots of sweets for us to sample.

An amazing example of chemical sensing occurs in many moths. Their **antennae** are covered with very sensitive hairs that can detect tiny concentrations of chemicals, especially those broadcast by female moths. Other creatures use their delicate chemoreceptors to sense the presence of organisms around them in situations we humans would not detect.

▶ **Touch:** the sense of touch depends upon **mechanoreceptors,** nerve cells that are sensitive to motion or pressure. Different parts of our body have different concentrations of these cells. Also, the depth beneath the skin varies. Parts of our bodies are more sensitive than other parts.

Some nerve cells transmit messages to the muscles via unconscious **reflexes**—remember when the doctor taps your knee and you kick your leg. Other touch responses are learned and may become habits.

Our nervous system is complicated and interesting to compare with other animals. All creatures need to interact with their environment. As humans, our ability to process the information received from our senses more than makes up for weaknesses we have when receiving signals.

## Vocabulary Words

antenna .................................... structure that transmits or receives radio waves

arthropods ............................. invertebrates having jointed legs and segmented bodies with external skeletons

chemoreceptors .................... nerve endings that are sensitive to taste, scent, and other chemicals

compound eyes ..................... visual sensing system of some organisms consisting of multiple units

cones ........................................ light-sensitive receptor cells in the retina responsible for color vision

DNA code ................................ sequence or order of bases (adenine, guanine, thymine, and cytosine) along a strand of DNA

frequency ................................ number of occurrences per time unit

Jacobsen's organ ................... sensory organ found in some vertebrates, especially in snakes

mechanoreceptors ............... sensory receptors that respond to physical stimuli

primate species ..................... species including humans, monkeys, apes, and other species

reflexes .................................... automatic motor reactions to sensory signals

retina ....................................... light-sensitive layer of cells in the eye

rods .......................................... rod-shaped cells in the retina that react to dim light

tapetum ................................... reflective layer behind the retina of cats and other creatures that improves their night vision

taste buds ............................... groups of cells on the tongue used for detecting taste

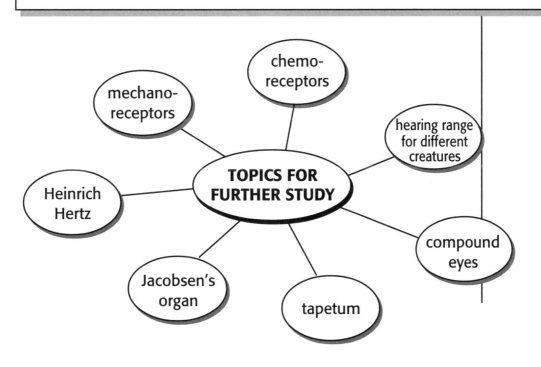

## Materials per Team

- variety of fruits, vegetables, and breads
- transparent containers with lids, or plastic bags
- magnifying lenses
- petri dishes with sterile agar or gelatin
- antibacterial soap and various disinfectants to test

# Germs

## *Wash Your Hands*

The process of decay is vital to life. Students can study the organisms that work to decompose other organisms. In this activity classroom teams can devise experiments and observe the decomposition process at work.

The first part of this activity is similar to the "Yeast and Molds" activity. Students can change some different variables this time around, or you may choose to move directly to the second part of this activity.

## Part 1

Tell students that today they will use several fruits, vegetables, and bread to observe the fungi that grow under various conditions. Give each team several plastic bags that will stay closed when sealed, and have them choose one kind of food for their experiment. Challenge them to find out what factors accelerate spoiling and what factors slow it down. Each team should generate a list of variables to test.

Next, have students place one sample of food in a bag and seal it closed. This is their control specimen. They should prepare their test variables from their list. Remind them to test only one variable per sealed bag. Here are some suggestions for test variables:

▶ Wet a piece of the food before sealing.

▶ Place one specimen in the sun or in a warm place.

▶ Add salt to a specimen.

▶ Place one in the refrigerator.

▶ Cook the food before bagging.

Once the samples are bagged, they should not be opened again.

Students should monitor their bagged specimens for a few days and record their observations. Provide magnifying lenses for close study.

Although there are many useful molds and fungi, these organisms responsible for decomposing can be dangerous to our health. Most foods must be kept clean and fresh so we can avoid ingesting harmful substances. Emphasize strongly that the specimen bags should be discarded after the experiment without being opened.

But it's not only food that can harbor tiny invaders. Keeping our hands clean is important for avoiding contaminating our food also.

## Part 2

Provide each team with several petri dishes, available from science supply companies, or try borrowing from your colleagues in middle and high school science. Fill clean petri dishes with a clear gelatin prepared according to package instructions and cover.

The idea is to contaminate the material in the petri dishes with fingers of varying cleanliness and compare the resulting growth of bacteria. Have students come up with the variables again. They can wash hands with different soaps, swab their fingers with alcohol rubs, use warm water washing only, wipe their fingers with a *very* mild bleach solution, and so on. Each student should test only one variable. Then have each student quickly wipe his or her finger across the gelatin in one petri dish and replace the cover. Leave some dishes untouched as controls.

Leave the dishes in a warm place overnight and check the next day to see if colonies of bacteria are growing. Then students can apply a disinfectant to some of the colonies to see if it kills them. Once again, use care not to touch the organisms and wash carefully after handling. Have students wash up after seeing the results of their experiments.

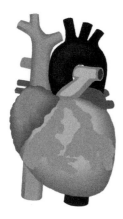

## *R E A D I N G:*
# Germit Theory

I n the nineteenth century the idea that people got sick when tiny foreign substances entered their bodies gradually became accepted. This **germ theory of disease** led to an emphasis on cleanliness and **sanitation.** When scientists discovered that the cause of many illnesses was invasion from outside the body, other scientists developed techniques to avoid contamination.

Ignatz Semelweiss (1818–1865) of Hungary suggested in 1847 that surgeons wash their hands to avoid spreading disease. Imagine a doctor *not* doing that nowadays! Joseph Lister (1827–1912) used **carbolic acid,** also called *phenol,* to cleanse patients' wounds, allowing the body to heal. The acid acted as an **antiseptic.** John Snow (1813–1858) traced the dreaded disease **cholera** to the water supply, and better sanitation became the rule.

Way back in the sixteenth century, P. A. Paracelsus (1493–1541) of Switzerland used strong chemicals, including mercury, to cure diseases. Mercury is a poison, and it was not uncommon for treatments to be as dangerous as the illnesses. Things changed dramatically in the nineteenth century with the gradual acceptance of the germ theory.

The **spontaneous generation theory** claimed that life could begin wherever conditions were favorable. Living things could appear without parent organisms or organisms moving in from elsewhere. In a suitable habitat (like aging food), life would arise on its own.

Louis Pasteur (1822–1895) was a French scientist who made several important discoveries about **crystals** early in his career. But he is most famous for his work in life sciences. He studied organisms responsible for **fermentation,** particularly **yeast.** Fermentation is a chemical process in which complex **molecules** are broken down and changed into simpler ones. Pasteur believed that the organisms causing the fermenting either arrived through the air, were added by people, or were present before the process began.

To refute the spontaneous generation theory, Pasteur used a **swan-necked flask** in his experimenting. He carefully sterilized liquids and kept air out of them with the neck of the flask blocked. He proved that the organisms did not arise out of the liquid. When he opened the neck of the flask, tiny forms of life could enter and begin to grow. Not everyone accepted his claim, but Pasteur developed heat treatments to rid food products of harmful **microbes.** Food spoilage was reduced, and the process was named **pasteurization** in his honor.

The germ theory of disease was hard to prove because the invading **microorganisms** were too small to detect, even with microscopes. Jakob Henle (1809–1885) and Robert Koch (1843–1910) were two German scientists who contributed to the germ theory. Henle formed a hypothesis about germs in 1840. Koch developed ways to grow bacteria and also discovered the cause of **tuberculosis.**

Pasteur discovered important traits about **bacteria,** some of the single-celled germs. Some bacteria are responsible for serious diseases. Pasteur's experiments taught him that heat can kill bacteria. He also learned to weaken bacteria in a laboratory.

When a weakened form of bacteria is injected into an animal or a person, defenses in the body mobilize and go on alert to repel attacks by stronger bacteria. This method of protection is called **vaccination.** Edward Jenner (1749–1823) had developed a **smallpox** vaccine in the 1790s, using a naturally weakened strain of the disease. Thanks to Pasteur, scientists learned to produce other vaccines.

Louis Pasteur's brilliant experimental work fighting infectious diseases and eliminating harmful organisms from food saved many lives. The shift in belief from spontaneous generation to the germ theory opened new fields of study in biology and medicine. It changed the way diseases are studied, attacked, and, if possible, controlled.

## Vocabulary Words

| | |
|---|---|
| antiseptic | chemical that inhibits or destroys microorganisms |
| bacteria | group of organisms with genetic material that is not contained in a nucleus |
| carbolic acid (phenol) | an alcohol sometimes diluted in water and used as a disinfectant |
| cholera | an infectious and often fatal disease caused by bacteria that affect the intestines |
| crystal | arrangement of matter in which the molecules are aligned in a regular, repeating structure |

# Vocabulary Words *(continued)*

fermentation ............................ process performed without oxygen, breaking down compounds into simpler forms, usually producing alcohol

germ theory of disease ........ theory that diseases are spread by "germs," or small organisms

microbe ................................... tiny life-form

microorganism ....................... an organism too small to be seen by the naked eye; microscopic

molecule ................................. two or more atoms bonded together

pasteurization ........................ process of destroying dangerous disease-carrying agents in liquid by heating; sterilization

sanitation ............................... behaviors designed for public cleanliness and health; prevention of dirty and harmful buildups

smallpox ................................. viral disease in which pimples form on the body

spontaneous generation theory ..................................... idea that organisms may arise from non-living things, sometimes called *abiogenesis*

swan-necked flask ............... laboratory flask with thin, curving neck

tuberculosis ............................ a disease causing lesions in the lungs

vaccination ............................. inoculation of a killed or weakened microorganism into people to protect them from infection

yeast ....................................... certain unicellular fungi

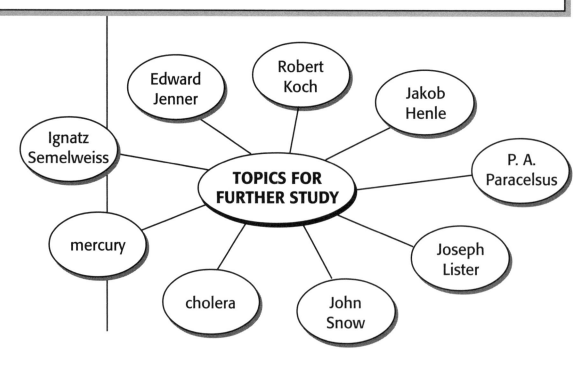

# Symmetry

## *Two Sides to the Story*

In "The Senses" activity, students experienced how having two eyes and two ears helps determine distance and direction. Other parts of our body are matched pairs, arms, legs, hands, feet, and so on. Humans, like many other animals, are quite symmetrical along our long axis. But how symmetrical are people?

### Materials per Team

- photos of students or magazine and newspaper photos of people
- small mirrors
- scissors
- calculators

---

## Part 1

Before beginning the activity, teach or review the concept of symmetry. In simple terms, something is symmetrical if it can be divided by a line and each side of the object is a mirror image of the other.

For the first part of this activity, you need photographs of people's faces looking directly into the camera. If you take photos of the students themselves, print a double set or make photocopies. Otherwise, cut out and/or photocopy magazine photos in which the subject faces straight ahead.

Hold a mirror directly at the midpoint of the photo. Angle it so that one side of the photo reflects in the mirror and produces a full face image. Chances are that the result will look strange. Alternate looking at each side of the face reflected in the mirror and notice how different they are.

Have students cut the photos down the middle and place each half on separate parts of a piece of paper. Ask them to try to supply the missing half of the face, either by sketching a mirror image or trying to draw it. Attempting that difficult task will emphasize the asymmetry in human faces. And don't overlook the opportunities to hunt for examples of symmetry in the classroom and beyond.

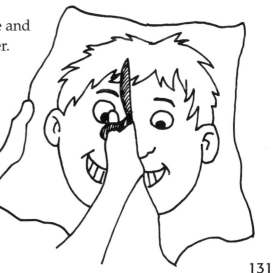

---

## Part 2

---

The second part of the activity engages students in some original research. What can they learn about dominant handedness? What activities are most likely to be done with the same side of the body as the dominant hand?

Brainstorm some suggestions and let each team of students devise a set of questions and tests to learn which side of the body they favor more often. For instance, if a person is left-handed, will he or she likely favor the left foot and left eye also? Generate some data using class members, and perhaps their families or other classes, as subjects.

To test for dominant eye, hand a person a cardboard core from a paper towel roll and see which eye they use to look through it. Ask students to invent other tests for their research. For example, which foot does a person use to kick a ball or push a scooter? Do we begin climbing stairs with the same foot every time?

Have students record data in a table. When they analyze data, students can practice many mathematical skills, including percentage and graphing. Finding information from wider studies will let them see how their class stands compared to general standards—for example, what is the average percentage of left-handed people in the class? In their grade level or school? In the general population?

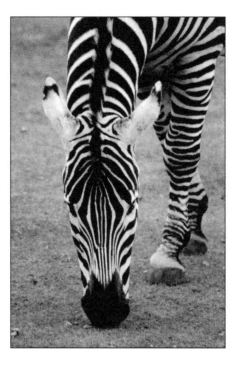

Most animals display symmetry. They may be divided into mirror halves along a line drawn on their bodies. (courtesy of Luis Rock, stock.xchng.com)

# READING:
# *Right Side to Left Side*

If you imagine a line down the middle of a person from top to bottom, the left side and right side will look pretty similar. Two eyes, two ears, two arms, two legs . . . you can tell that we are **symmetrical** to a large degree. Meanwhile, drawing a line *across* a person does not result in this almost-mirror image.

Think about some different animals; are they also **bilaterally symmetrical**? In other words, do they have identical parts on both sides of their bodies? What is helpful about this arrangement? For one thing, when you use your senses to receive the world's messages, you can usually tell which direction they are coming from.

The largest part of our brain, the **cerebrum,** has two distinctly separate **hemispheres.** If you look at a diagram of the brain, you will notice the major connection between the brain halves, the **corpus callosum.** Researchers continue to study the division of labor in our brain—which jobs are assigned to one hemisphere and which jobs are shared.

This research relies somewhat on **split brain research,** studying what happens when the corpus callosum is severed surgically. Many surprising facts about how we process information have been revealed. Michael Gazzaniga (b. 1939) reported from his research that the two halves of the brain can function independently. Both halves do many similar jobs, such as sending signals to our muscles.

Have you heard people talk about left brain and right brain thinking? Those terms refer to different kinds of brain activity. For example, playing music and writing a story are both creative activities, but the skills necessary to do each are quite a bit different. It makes sense that different sections of the brain will be at work.

Pierre-Paul Broca (1824–1880) discovered the first connection between a specific part of the brain and its function. During an **autopsy,** Broca discovered damage to the brain of a man who had lost the ability to speak. Walter Hess (1881–1973) later used electricity to stimulate parts of the brains of animals to find out what certain areas did.

You can do some of your own research to learn which side of the body people use more. How many people in your class are left-handed and how many are right handed? The hand used most often by a person is called the *dominant* hand. Does everyone throw and kick a ball from the same side of their bodies, or can you be right-handed and left-footed? Think about eye dominance. Is that trait always connected to hand dominance? When you present your research after collecting data, can you find a **correlation** among various types of dominance?

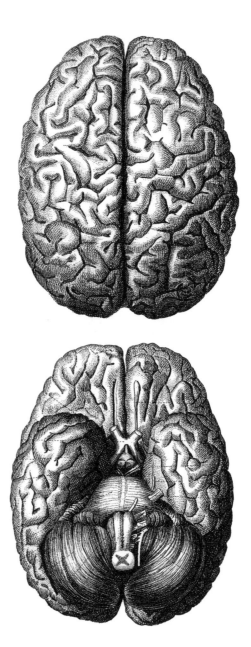

The cerebrum is composed of two connected hemispheres.

## Vocabulary Words

| | |
|---|---|
| autopsy | surgical investigation of a dead body to determine cause of death |
| bilaterally symmetrical | symmetry around two sides |
| cerebrum | largest portion of human brain |
| corpus callosum | in the brain, connecting strip between the two hemi-spheres of the cerebrum |
| correlation | mathematical relationship between two or more changing quantities |
| hemisphere | half of a sphere |
| split brain research | studies done with patients whose connections between the two halves of the brain have been severed |
| symmetrical | identical arrangement on opposite sides of a dividing line |

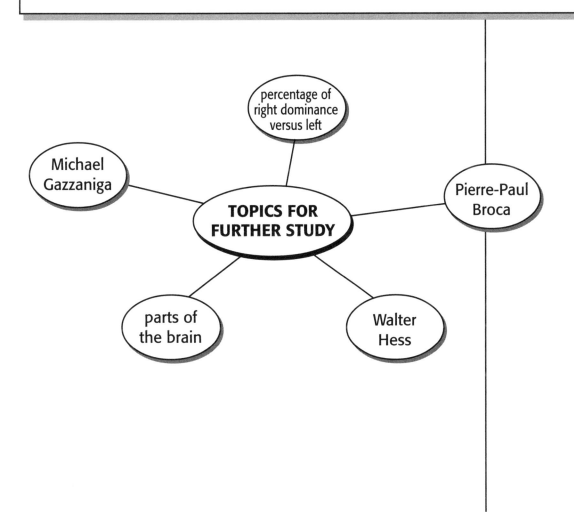

# Bibliography

Gribben, John. *Almost Everyone's Guide to Science.* New Haven: Yale University Press, 1999.

Gribben, John. *The Scientists.* New York: Random House, 2002.

Hazen, Robert M., and James Trefil. *Science Matters.* New York, Doubleday, 1991.

Hellemans, Alexander, and Bryan Bunch. *The Timetables of Science.* New York: Simon and Schuster, 1988.

Mayr, Ernst. *This Is Biology.* Cambridge, MA: Harvard University Press, 1997.

Meadows, Jack. *The Great Scientists.* New York: Oxford University Press, 1996.

Panek, Richard. *Seeing and Believing.* New York: Penguin Books, 1998.

Purves, William K., and Gordon H. Orians. *Life: The Science of Biology.* Sunderland, MA: Sinauer Associates, 1983.

Simmons, John. *The Scientific 100.* Secaucus, NJ: Citadel Press, 1996.

Timberlake, Karen. *Chemistry,* 4th ed. New York: HarperCollins, 1988.

Wallace, Roberta A. *Biology: The World of Life,* 5th ed. New York: HarperCollins, 1990.

Wilson, Edward O. *The Diversity of Life.* New York: W. W. Norton, 1992.

# Glossary

acid .............................................. sour-tasting liquid containing hydrogen ions (H⁺)

adaptation ................................... a trait of a species that helps it survive; behavioral change

aerobic ........................................ process requiring oxygen

alcohol ......................................... group of organic compounds containing carbon, hydrogen, and oxygen atoms (OH)

allergy .......................................... physical reaction to a substance in the environment

amino acid ................................... molecules crucial to the formation of DNA.

anaerobic ..................................... describes a process that does not require oxygen

anatomy ....................................... the structure of an organism

antenna ........................................ structure that transmits or receives radio waves

antibiotic ...................................... substance that kills bacteria

antidote ........................................ something that stops the action of a poison

antiseptic ...................................... chemical that inhibits or destroys microorganisms

Archaea ........................................ group of single-celled organisms classified as prokaryotes (cells without nuclei) along with bacteria. Many forms live under extreme conditions.

arms race ...................................... an escalation or weapons competition. An advance by one side triggers a buildup reaction from the other, prompting even more weapon stockpiling by the first side, and so on.

artery ........................................... blood vessel carrying blood away from the heart

arthropods .................................... invertebrates having jointed legs and segmented bodies with external skeletons

autopsy ........................................ surgical investigation of a dead body to determine cause of death

bacteria ........................................ group of organisms with genetic material that is not contained in a nucleus

base .............................................. substance that neutralizes acids, removing the hydrogen ions to form water

bilaterally symmetrical ................... symmetry around two sides

binoculars ..................................... optical device used for magnifying objects at distances. Binoculars consist of two telescope-like tubes, one for each eye.

binomial ....................................... having two names. In taxonomy, a two-part name consists of genus and species.

blood clotting ............................... massing of blood cells, designed to stop loss of blood from a wound

blood pressure .............................. force of the blood within the arteries

botanist ........................................ plant scientists

botany .......................................... the study of plants

cancer ........................................... general name for invasive malignant growths of cells

capillary ........................................ smallest blood vessels, carrying blood between arteries and veins

carbohydrates ............................... group of organic compounds composed of carbon, hydrogen, and oxygen atoms

carbolic acid (phenol) ................... an alcohol sometimes diluted in water and used as a disinfectant

carbon dioxide (CO₂) ..................... common gas composed of molecules with one carbon atom and two oxygen atoms

cell ............................................... the smallest, microscopic-sized unit of organisms

cellulose ....................................... carbohydrate used by plants as a structural material

cerebrum ...................................... largest portion of human brain

chemistry ...................................... science of the structure and properties of matter

chemoreceptors ............................ nerve endings that are sensitive to taste, scent, and other chemicals

chlorophyll .................................... green pigments that trap light for photosynthesis

chloroplast .................................... organelle containing chlorophyll that functions in the performance of photosynthesis

cholera ......................................... an infectious and often fatal disease caused by bacteria that affect the intestines

chromosomes ............................... structures in cells that contain genetic information

classification ................................. system of grouping organisms based on their similar characteristics

codon ........................................... a triplet of nucleotides in a DNA molecule that spells out or specifies a particular amino acid

compound eyes ............................. visual sensing system of some organisms consisting of multiple units

cones ............................................ light-sensitive receptor cells in the retina responsible for color vision coniferous .............................................. referring to a cone-bearing tree

controversial ......................................... describes an issue or idea that causes a dispute

controversy ........................................... dispute; difference of opinion

cork....................................................... plant tissue in outer layer of stem that prevents drying out. The cork of certain oak species is used for bottle stoppers and other applications.

corpus callosum .................................... in the brain, connecting strip between the two hemispheres of the cerebrum

correlation ............................................. mathematical relationship between two or more changing quantities

cosmology ............................................ branch of science dealing with the creation

crystal.................................................... arrangement of matter in which the molecules are aligned in a regular, repeating structure

diffraction ............................................. bending of waves around an obstacle

dissection .............................................. to cut apart for study

DNA...................................................... deoxyribonucleic acid; a double twisted helix inside a cell that contains the genetic code for that organism

DNA code .............................................. sequence or order of bases (adenine, guanine, thymine, and cytosine) along a strand of DNA

drought.................................................. prolonged period without rain

ecology.................................................. study of the relationships among organisms and the environment

element ................................................. substance composed of one type of atom

emit ...................................................... to send out or discharge

endangered species .............................. species in danger of becoming extinct

environment........................................... an organism's surroundings

Eukarya ................................................. set of organisms whose cells have a definite nucleus

eukaryotic ............................................. cells in which the genetic material is inside a nucleus

evolution............................................... the theory that genetic changes from generation to generation over time cause species to change gradually

evolve ................................................... act of changing gradually

experiment............................................. a controlled test of a hypothesis

fermentation ......................................... process performed without oxygen, breaking down compounds into simpler forms, usually producing alcohol

fire management.................................... technique in which controlled fires are allowed to burn to maintain the health of an ecosystem

flower.................................................... reproductive structure of a seed-producing plant; blossom

fossil ..................................................... the remains of ancient organisms

frequency .............................................. number of occurrences per time unit

fungus (pl., *fungi*) ............................... organism from the kingdom Fungi. This kingdom includes mushrooms, yeast, molds, and other organisms.

genes .................................................... sections of DNA that produce a trait in an organism

genetic.................................................. affecting or affected by genes; hereditary

genetically modified
(GM) species ........................................ organisms whose genetic material has been deliberately altered

genetics ................................................ the study of heredity

genus .................................................... category of organisms into which related species are grouped. Each type of organism has a scientific name consisting of its genus and species.

germ theory of disease ................... theory that diseases are spread by "germs," or small organisms

germinate .............................................. sprout

glucose.................................................. common sugar composed of molecules of carbon, hydrogen, and oxygen atoms

heart rate .............................................. frequency of "beats" or contractions of the heart

helix ...................................................... three-dimensional spiral

hemisphere ........................................... half of a sphere

hepatitis ................................................ disease of the liver

heredity ................................................. process by which traits are passed from one generation to another through genetic information

hierarchy................................................ the order of classification determined by how broad or narrow the groupings are

hormone................................................ chemical produced by an organism that causes an effect on another part of its body

hybrid.................................................... offspring produced by parents who are genetically different

hypothesis ............................................. explanation or theory

inheritance ............................................ characteristics transmitted by genetic material

intravenous ........................................... in a vein

Jacobsen's organ ............................... sensory organ found in some vertebrates, especially in snakes

joints ..................................................... connection between bones or parts of animal bodies that permit movement

kingdom ................................................ largest (most inclusive) category used in common method of classifying organisms

leaves .................................................... green structures of plants attached to stems, functioning primarily in photosynthesis

Linnaean classification ..................... system for categorizing organisms using groupings from the most general to a specific binary (two-part) name for each species

liver ....................................................... the organ in the body that works to form blood and to metabolize food

maggot ................................................. general name for the larval stage of some insect species, especially flies

marrow .................................................. soft tissue inside bone

mechanoreceptors ............................ sensory receptors that respond to physical stimuli

medieval ............................................... referring to the Middle Ages, an historical period in European history often considered approximately A.D. 500–1500

microbe ................................................ tiny life-form

microorganism .................................... an organism too small to be seen by the naked eye; microscopic

Middle Ages ........................................ historical period in Europe, approximately the years A.D. 500–1500

misconception ..................................... a mistake in meaning; the state of having the wrong idea about something

mnemonics ........................................... devices, symbols, reminders, and so on, that can be used to help remember something

mold ...................................................... name for certain organisms in the Fungi phylum

molecular .............................................. referring to individual molecules

molecule ............................................... two or more atoms bonded together

monastery ............................................ home for religious people

monerans .............................................. bacteria, organisms in the kingdom Monera

mutation ............................................... permanent change in the genes of an organism

mythical ................................................ from a story, usually about a hero or supernatural creature

natural selection ................................ the survival of individuals whose characteristics are advantageous for their environment and elimination of those individuals who do not succeed

naturalist ............................................... scientist who studies natural objects and organisms

nectar .................................................... sweet liquid secreted by flowers

nitrogen ................................................ gas that makes up almost 80% of the atmosphere

non-native plant species ................. species that has been transported by human activity to a part of the world where it did not previously exist naturally

nucleotide ............................................ segment of DNA consisting of a base, a sugar, and a phosphate

nucleus .................................................. in a cell, the part separated by a membrane and containing most of the cell's DNA. In an atom, a positively charged mass containing protons and neutrons.

nutrition ............................................... process or study of how organisms use food

offspring ............................................... descendants of organisms—for example, the children of human parents

oxygen .................................................. gas that makes up about 21% of Earth's atmosphere

pasteurization ..................................... process of destroying dangerous disease-carrying agents in liquid by heating; sterilization

penicillin ............................................... antibiotic derived from the penicillium mold

perspiration ......................................... salty liquid released through the skin; "sweat"

phlogiston ............................................ non-existent substance that early chemists mistakenly believed to be released in the form of flame when things burned

photosynthesis .................................... process by which plants convert carbon dioxide and water into carbohydrates using energy from the sun

physiology ............................................ study of vital life processes

pigmented ............................................ colored

pioneering species ............................. hardy organisms that first colonize an area after a drastic change

pollen .................................................... material produced by anthers of flowers that is the male element in fertilization

primate species ................................... species including humans, monkeys, apes, and other species

prokaryotic .......................................... cells in which the genetic material is not inside a nucleus

protein .................................................. specific chain of amino acids that is vital to enzymes and cells

protists .................................................. organisms from the kingdom Protista, a group that includes mostly one-celled organisms with a nucleus

pulse ................................... the throbbing of blood vessels caused the beating of the heart

rain forest ........................... region of the Earth characterized by high annual precipitation and year-round green forest

red blood cell..................... a cell that transports oxygen in the body

reflexes ............................... automatic motor reactions to sensory signals

renaissance......................... a revival, a reawakening of cultural achievement

resin .................................... substance produced by plants (and in modern times by human engineers) used to make a variety of products

respiration .......................... breathing, or in cells, the use of oxygen

retina................................... light-sensitive layer of cells in the eye

rheostat............................... an electrical resistor that can regulate current in a circuit. Dimmer switches are a type of rheostat.

RNA ..................................... ribonucleic acid, a chemical active in transferring genetic information and building proteins

rods ..................................... rod-shaped cells in the retina that react to dim light

roots .................................... part of a plant that anchors it, helps absorption, and may store food

sanitation ............................ behaviors designed for public cleanliness and health; prevention of dirty and harmful buildups

seed ..................................... fertilized ovule of a plant, capable of growing into a new plant under favorable conditions

smallpox .............................. viral disease in which pimples form on the body

sociobiology ....................... the analysis of human behavior from the viewpoint of evolution theory

species ................................ group of organisms that is considered one type. Generally, organisms within a species can breed among themselves.

sphere ................................. solid round figure

split brain research ........... studies done with patients whose connections between the two halves of the brain have been severed

spontaneous
generation theory ............. idea that organisms may arise from non-living things, sometimes called *abiogenesis*

stability ............................... steadiness; resistance to change

starch .................................. name for molecules of carbohydrates abundant in plants

static ................................... unchanging

stem .................................... stalk, slender connecting part of a plant supporting a leaf or flower

succession (plant) ............. orderly sequence of change of species occupying an environment over time

sugar ................................... general term for some large carbohydrates

swan-necked flask............. laboratory flask with thin, curving neck

symmetrical ........................ identical arrangement on opposite sides of a dividing line

tapetum ............................... reflective layer behind the retina of cats and other creatures that improves their night vision

taste buds ........................... groups of cells on the tongue used for detecting taste

taxonomy ............................ science of classifying organisms

transfusion .......................... injection of blood or plasma into the bloodstream

tuberculosis ........................ a disease causing lesions in the lungs

vaccination .......................... inoculation of a killed or weakened microorganism into people to protect them from infection

variable................................ changing. In scientific experiments, something a researcher changes to collect data about its effect.

variation .............................. difference or change in form or structure from unusual type

vein...................................... in animals, a blood vessel carrying blood toward the heart; in plants, a vascular structure forming the framework of a leaf

virus ..................................... tiny parasite composed of DNA or RNA and protein

volcanic ............................... from volcanoes, or from rock formed by volcanoes

volcano................................ an opening in the Earth's crust where molten rock and gasses erupt

white blood cells ............... blood cells that fight and remove foreign bodies in the blood

X rays .................................. radiation of a specific high-frequency wavelength in which the photons have high penetrating power and can pass through solid objects

yeast.................................... certain unicellular fungi